Person-Centred Ergonomics:
A Brantonian View of Human Factors

Paul Branton (1916–1990)

Person-Centred Ergonomics:
A Brantonian View of Human Factors

Edited by
David J Oborne
Rene Branton
Fernando Leal
Pat Shipley
Tom Stewart

Taylor & Francis
London • Washington, DC

UK   Taylor & Francis Ltd, 4 John St, London WC1N 2ET

USA   Taylor & Francis Inc., 1900 Frost Road, Suite 101, Bristol, PA 19007

British Library Cataloguing in Publication Data

Person-Centred Ergonomics
A catalogue record for this book is available from the British Library

ISBN 0–74840–0051–6

Library of Congress Cataloging-in-Publication Data are available

Cover design by John Leath
Typeset by RGM Typesetting Services, Southport

Printed in Great Britain by Burgess Science Press, Basingstoke, on paper which has a specified pH value on final paper manufacture of not less than 7.5 and therefore 'acid free'.

Contents

Ed. note. Reference to Paul Branton's publications are given in square brackets throughout Chapters 1–5. The second in these references indicates the page number. For example, Branton [40, 15] indicates page 15 of reference 40 in the bibliography at the end of this book.

Preface

It is in the nature of a science that it continually changes over time. New thoughts, ideas, concepts and studies help to shift the basic tenets of it in different directions – sometimes with large leaps, but most often by small (sometimes imperceptible) steps. However the change arises, though, change it will.

Despite its relative youth amongst the scientific disciplines, ergonomics has undergone a number of changes throughout its development. In the early decades of this century, although not called 'ergonomics', the science of work mainly considered factors such as fatigue and industrial health. In the 1930s and 1940s, the subject developed through its 'knobs and dials' era. While in the 1960s and 1970s, the quality and quantity of information transmission between 'man' and machine became of prime importance – albeit with an 'eye' still on the knobs and dials as well as on fatigue and occupational health. The rise in significance of human-computer interaction in the 1970s and 1980s, and the mechanistic models which the computer analogy naturally engendered, strengthened this perception of the individual being a major – albeit error-prone – actor within a closed information loop.

Despite having received his formal psychological and less formal ergonomics training during the 1950s and 1960s, Paul Branton rarely succumbed to the contemporary views of the discipline. His philosophical background and natural humanity led him to realise that, far from being sources of error in a system, people at work bring unique characteristics to the system which no machine would ever hope to match. They have a sense of responsibility in their work, a quality of flexibility and adaptability, and the ability to predict events and courses of action for the future.

By recognising such characteristics, Branton brought to his ergonomics practice and writings a far more 'person-centred' view than was (and, to a certain extent, still is) currently fashionable.

This book fulfils two aims. First, to enable the ergonomics community to read more widely Paul Branton's writings on the theme of person-centred ergonomics. Second, to explain and to develop the person-centred approach to ergonomics which was central to the Brantonian View. Enlightened ergonomists who begin to understand the view are likely to develop more useful and usable systems for people at work.

Many argue that ergonomics is currently at another of its developmental crossroads. The mechanistic, information-centred, view of people at work has begun to fail us when we deal with systems which are more complex than the simple design of individual components within a system. A shift in direction to considering the emotional, psychological and almost philosophical make-up of the person at work begins to offer a viable alternative. Thus, a third aim of this book is to help the subject shift its direction – to a Brantonian View.

D.J. Oborne
R. Branton
F. Leal
P. Shipley
T. Stewart

May 1992

Part I
The Brantonian View

Chapter 1

Person-Centred Ergonomics

Ergonomics is a young discipline, still in its early stages of development. Indeed, it is so young that its birthdate can be accurately pin-pointed: 12 July 1949. On this date, interested individuals congregated at the Admiralty in London with the aim of developing a new science of work (Murrell, 1980). The discipline which eventually emerged sought to amalgamate the knowledge and philosophies of such diverse subjects as psychology, biology, and engineering in order to understand better the ways in which humans interact with their surroundings, and thus how the surroundings can be engineered to suit the humans.

The name 'ergonomics' was created as an amalgamation of two Greek words: *Ergos* (work) and *Nomos* (laws). As a discipline, it takes as its starting point the constitution of individuals (psychological, biological, mechanical) within the working situation and how people interact with their environment. Thus, by understanding the wide range of important facets of these fundamental characteristics, in particular how the individual uses them to act within his or her world, ergonomists seek to manipulate the environment so that it will better fit the individual's own ways of working and behaving.

This novel view, that the job should be designed to 'fit' the person, represented a radical departure from the more traditional approach, which is still central to much work design, in which such techniques as selection, training, or even coercion, are used to adapt human operators to their environment (i.e. 'fitting the man to the job').

Although examples of the ergonomics approach can be traced back to analyses of working conditions around the turn of the century, the main thrust for the development of the discipline occurred during the Second World War and as a result of events which followed. Extremely sophisticated, sometimes novel, equipment was often unable to be used, effectively or safely, because the abilities required of the human operator were greater than those which could be delivered. For example, information from military systems often arrived too quickly to be processed at a cognitive level, visual or auditory displays were provided which could not be processed fast enough, the social and physical environments in which people were asked to fight were too extreme for effective work, and so on. An ergonomics approach, although perhaps not recognised as such at the time, often led to improved efficiency and an ability to use effectively the systems available. Thus, the need to understand what the human operator *could* accomplish and to design the system to accommodate these abilities led ultimately to the Admiralty meeting in 1949, and to the emergence of ergonomics as an important science of work.

The Man-machine system

In its traditional form, ergonomics seeks to maximise safety, efficiency and comfort by designing the 'machine' (or any aspect of the environment with which the person has to interact) to match the operator's abilities. By creating this formal link, a relationship is established between the two components ('man' and 'machine') so that the machine presents information to the operator via its displays and the operator's sensory apparatus, to which a response is generally made in some way – for example, the operator may use his or her limbs to operate controls which in turn alter the machine's state. In this way a closed-loop, error-correcting, information-transmitting system is developed such that deviations from the desired 'state' (however that state is defined) are displayed by the system, interpreted by the operator, and corrected using controls. Such deviations may arise from forces which are external to the system and from activities within it – either from the system itself or from the operator. Naturally, any complex working situation will be composed of a variety of different man-machine loops of this nature. For traditional ergonomics, however, the single loop system remains the ideal unit of analysis.

Taking this conceptualisation, therefore, ergonomics can be said to be an amalgam of physiology, anatomy and medicine as one branch; physiological and experimental psychology as another; and physics and engineering as a third. The biological sciences provide information about the body's structure: the operator's physical capabilities and limitations; body dimensions; how much can be lifted; the physical pressures that can be endured, etc. Physiological psychology deals with ways in which the brain and nervous system function and determine behaviour; while experimental psychologists study the basic ways in which individuals use their bodies to behave, perceive, learn, remember, control movements, etc. Finally, physics and engineering provide similar information about the machine and the physical environment with which the operator has to contend. Interestingly, this list of discipline-based contributors does not feature the social side of science – despite the fact that, in many ways, it is the person's behaviour which defines the system's effectiveness.

A person-centred ergonomics

The traditional view of ergonomics is thus one which argues that the individual and the working system must operate in close harmony for efficient operation of that system. However, it is the individual who controls the system, who operates it, steers its course and monitors its activities. In doing so, it is the operator who has goals and wishes and who can change the system through abilities and whims. It naturally follows, therefore, that the system must be designed with the operator's capabilities and requirements in mind if it is to be effective.

The approach, therefore, perceives the complete, working man-machine system as one in which action and communication flows in an almost symbiotic way between the operator and the system; the one depends on the other for continued operation – almost for continued existence.

More recently, however, this view of the fundamental relationship has been questioned by an increasing number of authors as being too simplistic. Discussing the human-computer interaction domain, for example, Eason (1991) argues that this viewpoint 'as a form of conversation between different kinds of participants' misses the rich complexity of the interaction. He points out that we interact with machines not merely to exchange messages but to engage in complex tasks in the real world. The man-machine interaction, then, has a meaning which is over and above that which can be expressed by simple, straightforward analyses of the component parts. This meaning is injected into the system by both the individual operator and the nature of the task and its outcomes.

Wisner (1989) takes the argument beyond just interacting with computers to encompass the full domain of ergonomics. He argues that what is specific to ergonomics, as well as to psychology, is that it should not try just to investigate specific 'properties of man'. Rather it should attempt to 'understand how man uses his own properties in terms of a story, his own story and that of humanity, the part of humanity to which he belongs'. Individual wishes and desires, motives and experiences, then, are brought to a working situation and must be understood when considering the 'fit'. Even such factors as social background and culture play an important role.

So the shift in ergonomics thinking which has emerged in recent years has been towards understanding more the nature of the individual within the system. The original concept of person and machine being almost equal partners is gradually giving way to a view which argues that the partners are not equal; that the person should play a more central role within a working system.

Thus the person-centred philosophy within ergonomics views the interaction as one which is controlled and dominated by the operator(s) in the system. In the course of their interactions with it, individuals bring to the system a collection of inherent strengths and weaknesses (from such factors as experiences, expectations, motivations, and so on) which themselves will interact with the system to change it. Such changes are usually to the good of the system, from the viewpoint of such criteria as efficiency and safety. Sometimes, though, the features which the individual brings to the situation will include variability, fallibility and maybe even perversity – any of which are more likely to lead to errors and inefficiency. Both facets of the person's input are emphasised in Wisner's words: that operators will use their 'own story' when being asked to control and manipulate events.

Put in another way, with others Branton [31] has emphasised that human operators in a system inevitably turn it from being a closed-loop system in which information flows from one component to the other with (in theory) maximum efficiency for correcting deviations within the system, to an open-loop system. Instead of a 'designed' error correcting system, then, the deviation corrections are effected by the operator on the basis of his or her 'mental model' of the system and its operation.

Thus the traditional ergonomics philosophy of a mechanistic relationship between an operator and a machine can be said to have misperceived a critical perspective of the system in which it is involved i.e. the operator and the attributes which he or she brings to the system in the wider sense. So, rather than concentrating simply on ways of improving the information flow between components within the system (at least the components which take account of the human operator), person-centred ergonomics takes as its central point the need to accommodate the human attributes which the person brings to the system. The emphasis is, thus, on accentuating the positive and reducing the negative effects of the individual's interactions.

Indeed, this missing perspective is especially important in the way in which it suggests *how* the ergonomics should be considered – particularly the point within the system at which an ergonomics intervention is made and how it is made. To take a very simple example – the design of a control for safe and efficient operation. By emphasising the closed-loop nature of the interaction and the need to *design the system* to fit the user, traditional ergonomics argues that the control design – the point of the ergonomics intervention – should be made from the viewpoint of the machine's requirements as they impinge on the operator's wishes and abilities. Certainly, the operator's abilities are important, but it is the system which defines the interaction and thus the design.

By viewing the interaction from the other side, however, from the viewpoint of the individual, the person-centred approach argues that the intervention into the closed-loop system should be made *at the level of the operator*. He or she is the component which is designed to activate the system and to maintain its efficient running. Thus the person-centred perception is one in which the operator and his or her abilities define the working system. So the goal is to create supportive dynamic environments which enable individuals to work at their most safe and effective levels; not just to design the environment to 'fit' the person in some static sense.

In many respects, the traditional and person-centred perspectives can be viewed as being variations in emphasis as to which are the most important components within a system. The traditional view emphasises the individual almost as being subordinate to the system; the person-centred view concentrates on the individual as a unique controller of the system. By taking such a position, however, the person-centred approach loses none of its thrust towards the ideal of creating working environments that will fit the abilities and requirements of human operators. Rather, as this book illustrates, potentially the effectiveness of ergonomics is strengthened considerably as a result of the change in emphasis.

The Brantonian View

The Brantonian view of ergonomics follows this more person-centred approach, with significant emphasis on the nature of the individual who is to do the controlling and regulating. Thus Branton argued for consideration of such features as the importance of individual behaviour within this equation, and how their view of the world will help people to adapt their behaviour in the light of events. As will be seen in later chapters, he argued that it is the effect of such human factors as autonomy, sense of responsibility and creativity that are brought by people to their jobs which may render an otherwise potentially inefficient and inflexible system more valuable. Branton, thus, took a humanistic view of ergonomics, mustered at the same time with all possible scientific rigour.

Paul Branton (1916–90) trained originally as a psychologist, having previously spent 20 years in industry and business, with an additional six years' war service in the Royal Navy. Before University, he was particularly concerned with understanding the bases of personal value judgements, and studied critically the philosophical foundations of ethical behaviour in general, and morals and politics in particular. This interest in the philosophical bases for understanding behaviour, particularly in the tradition of the Critical Philosophy, remained with him throughout his life and helped him to develop and maintain a specific person-centred view of ergonomics – the Brantonian View.

Following an Honours degree in Psychology from the University of Reading (UK), during which time he developed a strong interest in the physiological and anatomical bases of psychological events, Paul Branton moved to the Furniture Industry Research Association (FIRA), with a brief to investigate the comfort of sitting and seats as part of an ergonomics service to that industry. During his period at FIRA (1962–66), he further developed his thoughts about the nature of ergonomics and the prime importance of the individual within the system. In particular, he began to develop views about ways of integrating observable behaviours such as movement and posture with the less observable and more subjective behaviours, particularly 'comfort', in order to understand better, and fully meet, the needs of the user.

During his period at FIRA, he began to develop his theory of postural homoeostasis to explain individual comfort-seeking behaviour. He then moved to the MRC Industrial Psychology Research Unit where his interests shifted more towards global industrial problems and how psychological interventions and investigations may help to explain them. It was here that his interest in shiftwork and accidents began to take shape, leading particularly to development in his understanding of stress and decision making within individuals.

The final formal shift in his career occurred in 1969 when he joined British Rail as its Chief Ergonomist. Here he was free to develop his ideas in a specifically applied setting, as the following list of projects illustrates:

- Train driver's attentional states and heart rate studies
- Design of driving cabs and anthropometrics of drivers
- Display/control instruments
- Skill analysis of driver's tasks and route learning
- The design of signals to be observed; laboratory studies of vision
- Inspection problems: rail testing, freight vehicle examination, film inspection workload
- Signal box design
- Control rooms
- Design for safety: accident reporting procedures, safety of 'permanent way' men
- Perceived space within carriages
- Design of seats and comfort characteristics
- Timetable design
- Passenger flow
- Subjective response to vibration in vehicles.

Reviewing much of his work, it is clear that his interests had centred on a number of focal areas, particularly the nature of:

- skilled behaviour
- stress
- comfort, and
- decision making.

Indeed, Branton continued his research interest in the first two of these areas after he had retired from British Rail, when he became an Honorary Research Fellow at the Stress Research and Control Centre at Birkbeck College, London.

The above lists are by no means exhaustive but serve to illustrate the wide variety of applied ergonomics topics which Branton considered. Each was not studied just for its own sake, however. Instead, each was used to investigate the underlying processes involved in an individual's behaviour at work. To each area he brought the person-centred view of ergonomics; that one should consider the whole person within the system – psychologically, physiologically and *philosophically*.

The need to understand the individual's personal philosophy and psychology is central to the Brantonian View:

> My first praise of ergonomics is for the breadth of its possibilities; it constantly challenges me to consider the whole person. . .It was by sheer luck that I came to ergonomics early in my academic career. Previous experience in industry and in the war had turned me to practical matters, yet I always had a speculative streak in me. Ergonomics with its emphasis on concrete application, virtually forced me to compare theory with reality and adopt an entirely fresh approach to puzzling facts, revealed by actually observing people's behaviour. None of the then current theories offered adequate explanations and so the new angle of view led me almost inevitably to attempt constructing my own abstract frame of reference. When applied consistently it became a coherent account of purposivity in the actions of men and women, especially when at work. By purposivity I mean that our actions are determined by the future rather than by the past. Many things, like skilled behaviour, are not merely 'goal directed' but are

best explained by asserting that we possess a specific capacity to represent to ourselves internally our own future actions – and their consequences – *before* we actually carry them out. It is one of those puzzles that we manipulate events that have not yet actually happened; and we seem to do it quite successfully. [40, 2]

This quotation encapsulates the Brantonian (person-centred) view of ergonomics:

- that the complete individual needs to be considered within the system – not merely that part of the individual which is interacting with the system (the eyes, limbs, or cognitive structures, for example).
- that individuals have perceptions and views which extend beyond the limits of the present, the situation under consideration. Thus one should consider also the individual's philosophy; his or her 'theory of the world'.
- that people have values which must be recognised, understood and accommodated before the system can be designed properly for them.
- that people at work act in an experimental and scientific way, continually testing a situation (both consciously and sub-consciously) to explore possible outcomes and their effects. Thus a single man-machine system is not static; it is dynamic and constantly responding to inputs from the human operator.

A...model is proposed which takes the form of a strictly autonomously controlled person, necessarily possessing value standards and interests in social relations, perpetually seeking and evaluating information from the surroundings. The search varies in intensity, depending on arousal level...The model concerns the form of mental operations, their contents being material either taken immediately from the surrounding world or from stored, primarily emotive experiences. The explanation is purposive, rather than causal, as it is argued that the thoughts which determine behaviour are forecasts of future states of affairs and their consequences for the person, rather than past experiences in themselves of speculative origin. [35, 505]

As it stands, the person-centred approach represents a major shift in the way in which ergonomists should view the role of the operator within the working situation. As the above quotations emphasise, however, Paul Branton added two important inputs to the approach to develop the Brantonian View: from psychophysiology and from philosophy.

The psychophysiological input

In many respects the person-centred approach, which argues that ergonomists need to understand the nature of individuals and what they bring to a situation as human beings, begs an important set of questions. The approach lacks persuasive powers until it is able to indicate *how* such features might arise and assert their effects. Simply to argue that boredom is an important and unpredictable human characteristic, for example, begs the question of how it occurs, can be measured, and can possibly be reduced.

The traditional ergonomics approach would be able to provide a perfectly sensible response to such a question: it would urge ergonomists to investigate the task variables as precursors to boredom and to relate such requirements in a redesign of the job to match the operator's abilities. In contrast, on the face of it, the person-centred approach lacks this kind of precise strength by advocating, as it does, an understanding of the 'humanistic' needs of the individual. The Brantonian View, however, provides some tangible basis from which such a general view might proceed.

Branton's response to such questions often rested in the physiological domain. He frequently asserted that it should be possible to understand person-centred variables from a psychophysiological perspective. For example, train drivers' monotony and boredom could be viewed in terms of physiological adaptation and stimulus inhibition [11]; anaesthetists' stress could be considered to be the result of an accumulation of 'mini-panics' arising from a realisation that they, as well as their patients, had drifted off to sleep (though fortunately only momentarily) during long operations [24]; and seating comfort (actually discomfort) behaviour could be considered to be the result of physiological adaptations to postural instabilities [7].

The recurrent theme which runs through many of these cases is that of rhythmical variations in bodily functioning. As will be discussed in more detail later, Branton was much influenced by the fact that our bodies exhibit such rhythms in almost all of their functions – consciously and unconsciously. These fluctuations, he argued, can be perceived as being major influencers of our behaviour when interacting with working systems. If not taken note of in design they could lead to some of the errors which the operator might induce into the system or, more likely, to cause operators discomfort – though they are still likely to manage to maintain the system under safe and efficient control.

An important variation on this Brantonian theme is that the physiological and psychological make-up of humans is such that people learn to compensate autonomously for their biological, and other, 'weaknesses'. This perspective, then, stresses that the ergonomist's task is to design a supportive enough environment to facilitate such compensating behaviour.

The philosophical input

Some would argue that the foregoing 'principles' of the person-centred approach are little more than a mere realignment of emphasis of the basic principles of ergonomics, and to some extent they would be correct in their assertion. Ergonomics takes as its central tenet the importance of the individual within the system and of understanding the individual's needs and abilities when interacting with the system. This forms the basis for ergonomics design. However, the Brantonian view within the person-centred approach to ergonomics extends this thesis in three significant directions:

1. the need to consider the *complete* individual as argued above (traditional ergonomics rarely does).
2. the need to consider *purposes* rather than *causes* of actions (a purposive explanation of an event – in terms of anticipation and decision making – is likely to be more illuminating than a deterministic causal one – which will be couched in terms of events which have already occurred).
3. the need to understand the philosophical bases upon which individual behaviour within the system rests; how he or she conceptualises the system and its functions.

Thus, another input within the Brantonian approach is from philosophy:

> We begin by accepting Kant's transcendental argument that humans cannot know the 'real truth' – whatever it may be – with *absolute* certainty. For the possibility of 'knowing anything at all', we must therefore ultimately resort to sources of knowledge which lie *within* ourselves. Such knowledge as we have, we possess only as actually thinking beings who possess by their own effort whatever information they receive from the outside. [28, 14]

As the above quotation illustrates, Branton's philosophical tradition extends back to Kant. Leal illustrates later in this book that further investigation has shown that it can then be traced from Kant through to the German philosopher Leonard Nelson (1882–1927) and his twentieth-century pupil Grete Henry-Hermann.

> The Critical method, to which Nelson was committed, attracts me, as a psychologist, because of its applications to research methodology as well as to the formation of theories about functions of the human mind and body. I see great benefits for both psychology and philosophy if they could pool their present knowledge to concentrate on the problem of which human faculties and capabilities are really the most basic ones. Even today many theorists go back to Aristotle, Aquinas and Hume and adopt their views seemingly without taking account of how much more complex processes like perception and cognition are than these savants could possibly have known. As a result, anyone working in the fields of social science, in education, psychiatry or any of the other applied human sciences must be utterly confused by contradictory theories and thus in urgent need of up-to-date conceptual frameworks.

> Nelson, extending the Kantian tradition, pursued the argument that possession of knowledge of reality, necessity, causality and other notions connecting events, must come before experience. Indeed, a person must first have the capacity to turn single events into coherent experience. Nelson's special contribution concerned the linking of epistemology and psychology without either dominating the other: every metaphysical proposition must be matched by another statement, namely that this is psychologically possible. [28, 14]

The critical tradition, or 'critique' for short, is composed of two parts: a destructive or negative phase and then a constructive or positive one. The tradition is summed up well by a documented discussion between Nelson and one of his students:

> *Student*: I have come to the point when I must tell you that I do not believe a single word of your deduction of the moral law.
> *Nelson*: You are then on the right path.

The tradition of critical philosophy, therefore, is to approach the problem from a questioning viewpoint, asking the kinds of questions which we often do not think about consciously. These are questions about the individual's purpose, moral values, intentions, and so on. That is the destructive or analytic phase. The positive or synthetic phase is then to draw information from a variety of areas to begin to explain some of the questions.

> Critique can be contrasted with the usually practised dogmatic method when it comes to choice of research topics and criteria. If this raises the discussion to the level of meta-psychology, so be it! Whereas the initial choice of the questions to which answers are sought presents little problem to the dogmatist when he sets up his hypothesis to fit his theories, the discerning and heuristic researcher turns to critique and asks for the – as yet – hidden meaning of what he observes. Hidden, because no amount of 'hard data' can of themselves alone bring the insight into the connective ideas and relational abstractions between observed events which is needed to uncover new knowledge.

> . . .Critique is not theory. The relation of critique to explanation is that the former first seeks to establish what it is that needs explaining, whereas explanation consists of theories in the form of conjectures and abstractions about the phenomenon being studied. . .Critique. . .encourages the habitual focusing on what questions are to be asked before rushing out of the lab again for further evidence. [32, 2–3]

and again:

> To find a way out of this conceptual maze, one guiding thread may be offered to the researcher. It is to adopt an 'epistemic' strategy: to ask himself in the first place at each stage 'How do I come to know whether this or that statement by the operator is true?' 'What is the source of my knowledge?' 'How direct or indirect is my perception of the measurement?' 'How far is my conjecture based on analogy and how far does it penetrate to the "real thing" ? ' Having thus become conscious of their own inevitable bias, observers (and their readers) are better able to speculate on their subjects' knowledge, values and actual purposes. [35, 505]

In many respects, the same processes of analysing the system's components and then drawing on information from a wide variety of areas to bear on some kind of answer to the problem can be seen within ergonomics. By its integrative nature then, ergonomics lends itself well to the critical tradition – and vice versa.

From Kant to Branton

In his *Critique of Pure Reason* (1781), Kant argues that rational psychology, i.e. non-empirical knowledge of the soul or mind, is impossible. Implicitly, there-fore, he asserts that empirical psychology *is* possible. Unfortunately, since there was no established science of psychology in his time he was unable to analyse critically the nature of these experiences and to follow up the implications of his theory. Critique, the branch of philosophy which Kant founded, can be viewed in the same way as psychology, insofar as its purpose is to analyse critically; it is thus a psychology as applied to the task at hand. Paul Branton was one of the latest of a long line of philosophers to argue this position.

Kant, however, was not content with that basic level of analysis. He felt that a *Critique of Practical Reason* was also needed, which he produced in 1786 and which posed the question: How is action (as opposed to knowledge) possible? Through this consideration, Kant hoped to develop metaphysical principles of rational action (or 'reasonable conduct' in Branton's terms) – principles for which psychologists and ergonomists still search when considering behaviours such as skills, decision making, and so on.

Kant's main preoccupation was with moral or ethical behaviour. Thus, when he spoke of (rational) action he meant moral action; such aspects as good deeds, resistance to temptation, and a virtuous disposition. On the surface at least, such moral considerations can be said to play only small roles in ergonomics endeavours and even ergonomics curricula. In relation to ergonomics, however, a Brantonian View of this philosophy paints a rather different picture as far as the moral imperative is concerned. Branton argued for a set of metaphysical principles of work which are not at all separable from moral considerations. On the contrary, he felt that moral considerations intrinsically belong to ordinary human work.

The question of moral behaviour, then, can be considered from both 'sides' of the work and design. On the one hand, human work involves what we may call normal responsible behaviour on the part of the human operator. From the beginning, the worker is part of a socio-technical system which embodies moral values and makes moral demands on all people having to do with it. Stress, for example, can sometimes be understood as the moral reaction of the responsible operator who realises that the standard of his or her performance has fallen or is being compromised in some way:

> To distinguish between 'beneficial' and 'harmful' stress, it may, for instance, be interesting to know the effects an operator's responsibility has on his own body. [40, 14]

and

> In field studies [of transport operators], brief episodes of absent-minded-ness have been found which normally recur cyclically at about 100 minute intervals, akin to the dream stages of REM sleep. When suddenly emerging from these lapses, 'mini-panics' have been reported. It is thought that the

sudden consciousness of having lost control over oneself, whilst being fully responsible for the lives of others, may be the major source of stress in transport. [35, Abstract]

Moral considerations are also important from the ergonomist's viewpoint. Thus we have a moral obligation to design the work place in ways that ensure the health and safety of workers and which are appropriate to them as human beings, that is, environments which are adapted to people's capacities and needs and which avoid unnecessary stress and misery.

The premise. . .is that the process control operator is in fact a responsible person even if he is not always treated consistently as such. If a man is to be responsible, he must be given the means for this and is not to be regarded merely as an automaton. [31, 1]

In whatever ways work is viewed, then, a variety of moral values are involved, each of which can affect the operator's behaviour and response to the situation. It is this argument which remains at the core of the Person-centred approach to ergonomics.

Summary

This chapter has outlined the philosophy of the Person-centred approach to ergonomics, as expressed through the views of Paul Branton (1916–90). Whereas all ergonomics considerations realise the prime importance of the human operator within the system, the person-centred approach extends the philosophy in two significant directions:

1. That the entry point for the ergonomics study must be from the viewpoint of the human being within the system rather than the system's requirements.
2. In doing so, the *whole* individual must be considered, rather than a part of him or her. This includes the individual's view of the system, purpose and responsibility within the system, and the dynamic nature of the interaction. In particular, the inner processes of individuals need to be studied as well as the traditionally observable behaviours, and full cognizance must be taken of the values of individuals within the system.

These considerations lead me to another virtuous aspect of the breadth and depth of Ergonomics: self-understood acceptance of multi-disciplinary connections, all within the frame of practical human sciences. A discipline which combines an interest in the environment with a search for inner consistency demands a holistic, rather than a limited, answer. If everyday life is all of one piece for a person, however chaotic it may seem, the ergonomist's descriptions and explanations must also be ever more inclusive and interactionist rather than reductionist. [40, 6]

Many of these aspects will be addressed through a thorough critical consideration of the *philosophy* of the working situation, which will help to develop methods and techniques for its study.

> My excuse for this philosophical digression is that it strengthens my confidence in a theoretical framework which may eventually take ergonomics – and other sciences – into new fields. It may even free one from barren forms of empiricism, to look beyond one's nose. Who knows, some ergonomists might even be tempted to speculate or philosophise mildly for themselves rather than meekly accept Descartes and Hume. [40, 10]

Altogether, the Brantonian View represents a major shift in emphasis in ergonomics thinking towards what Paul Branton described as a 'Design from the Man Out' approach. In this view, the human being is at the centre of the working situation, and we must understand the abilities, responsibilities and requirements which people bring to the situation, not just their foibles, in order to be able to deal adequately with the system. Many of the points addressed by a Brantonian view will be considered and expanded in the following chapters.

Chapter 2

Human values

By conceptualising the working system in such a way that the human operator is at the centre of events – as *the prime* component within the system – and by studying the system within this framework, the person-centred approach places considerable emphasis on the paramount needs of the individual and the features, both physical and behavioural, which he or she brings.

To describe it in its most basic form, the approach argues that human beings have special abilities and 'flexibilities'. When we understand the person we will begin to understand the system, and when we understand the person's variabilities we will be in a better position to adapt the system to accommodate and even to make best use of them. And when we can do this we should be better placed to design the system for maximum efficiency, safety, comfort, and so on.

Understanding the behaviour of people, however, is no simple task. Psychologists and philosophers have been trying to come to terms with the problem for many hundreds of years, so far without a great deal of success. Nevertheless, within a person-centred approach it is possible to identify principles and areas which play an important role in influencing the behaviour of people at work and which will have implications for the design of a system (particularly if we can relate the behavioural processes to underlying physiological ones). These areas specifically concern the 'values' which individuals bring to a system in order to act autonomously within it, and the nature of the system's features which will satisfy these values.

Purposivity

In most cases it is self evident that the operator at work performs his or her task with a sense of purpose. He or she has some reason for carrying out the action, some goal to attain: to throw a switch, illuminate a bulb, record a message, or whatever. Without such a purpose the action has no meaning as far as the system is concerned and can be classed as 'random' movements. No amount of ergonomic design will be able to accommodate such behaviours – at least not for any sensible reason other than to ensure that the unpurposeful movements do not damage the system by increasing the likelihood of inefficiency or accidents.

The person-centred approach, therefore, argues strongly that one of the primary features that the individual brings to the system is a sense of purpose, of action. When the purpose is understood it should be possible to begin to design the system in order to facilitate it.

To take one example of this approach: Wright (1986) suggests that a traveller who views an information display board at an airport does so with a preconceived model of the structure of the information in his or her 'mind'. For example, someone may want to access the information in terms of the flight number, the destination, or the time of departure. From their different starting points different individuals may then desire different features from the information: departure gate number, whether the plane is delayed, and so on. Whereas it would be impractical to design a separate airport display board for each individual traveller's expectations, it should be possible to determine the general purpose of the display board for use by the majority of passengers.

By understanding the purpose of the individual at the point of his or her entry into the system, therefore, ergonomists can begin to consider designs which will facilitate this aim, and allow the activities to be carried out most efficiently. Taking this thought a little further, in an unpublished paper in which he discussed the value of anthropometric data, Branton argued for a 'principle of function' to be introduced into human factors research.

> Activity and intention (or purpose) will determine the operator's postures and movements, and thus the actual envelope of working space required, at least as much as the equipment itself. [15, 9]

He continued by arguing that if activity and intention are determining factors in how the equipment will eventually be used, it is necessary to foresee these actualities to avoid misuse of the equipment. Thus data (in this case anthropometric data) should be obtained and applied with its functions in mind in order that the user's intentions regarding the system can be facilitated.

Unfortunately, however, it is not always possible to extract from the situation the full complement of purposes in which an individual engages when carrying out an action. Evidence from the field of expert system design, particularly of the difficulties involved in knowledge elicitation, has demonstrated that problem very well. Thus the very nature of skilled behaviour is such as to routinise many of the activities which compose it; as such they are carried out in an almost 'unconscious' way and so are difficult to verbalise:

> First . . . I interviewed hundreds of drivers and inspectors about 'what they really did when driving'. They eloquently described the route in terms of personal and unofficial cues, rather than in signals and other line-side equipment provided. But by far the most striking thing was that these men, some with 30–40 years experience and with high IQs, were quite unable to put into words how they drove and what they really did, and when they decided to do what they did. [40, 7]

Nevertheless, with care it is possible to extract in some form the relevant information from people. In this respect the important theoretical implication of this approach is that people are not just 'doers', they are thinkers too. They generate mental models about the world and impose into these structures the potential consequences of their actions. This involves both predictive and anticipatory abilities, both of which must be understood and accommodated within ergonomics designs.

As an example of this requirement, Branton has provided a running commentary of a train driver who carries out his tasks by continually developing and changing his working models of the task and its environment:

> As I cross the little bridge here, when I'm doing 80, with 8 coaches behind me, and I have to stop at Didcot, and I have plenty of time, and the weather is good and the lights with me, then I shut off the power here and I'll coast nicely to the station. Just before the station entrance I give her a 5lb rub on the brake before taking it off altogether and she'll stop about 4–5 yards from the signal post. I've done that many times.

As Branton explained the situation:

> Our interpretation of this relatively simple statement is that, over and above his watching for danger signals, the driver makes running computations of the effect of gradients, curves, signal aspects, etc., upon his subjective, intuitive solution of the total time/distance equation which forms the current mission. On perceiving changes in any of the variables, he re-calculates his subjective trajectory by a process closely akin to partial differentiation and approximation, and he thus minimises the difference (or error) between his forecast and the actual performance. [21, 178]

There is little evidence in this example of a simple 'doer' as far as the skill of train driving is concerned. Rather, the individual's purposivity involves considerable computations. Indeed, Branton felt that train drivers operate in a 'quasi-mathematical' mode, which suggests that some major re-evaluations should be made regarding the traditional perceptions of the skills involved!

> To attempt to describe them [the quasi-mathematics], we start from the insight by Bruner (1964) that humans construct models of their world by way of three information processing systems: through action, through imagery or through language, resulting in three different types of representations: enactive, iconic or symbolic. We assume that our driver's representations are in the enactive mode, because he is manifestly able to perform them but not so able to express his actions in language, let alone in numerical symbols. [21, 180]

We shall return to the nature of purposeful behaviour, and its implications for understanding and explaining the operator's intentions in skilled activity (which are often un*self*conscious ones), in the next chapter.

Anticipation and prediction

The existence of purposivity implies also some concept of prediction and anticipation; to decide to throw a switch, for example, and to perform the cognitive and muscular actions to accomplish the desired goal, one must have a fair idea of the likely outcomes. By implication, therefore, the concept of purposivity also suggests that the user will have some notion of the desired outcome, and thus some 'internal' model of the course of future events. This is an important linkage as far as person-centred ergonomics is concerned. Prediction, and the anticipation which arises from it, concern the operation of the system within the individual's perceived control, and will have major implications also for the nature and the quality of the information required for the task. Discussing the information obtained from train drivers, for example, Branton declared:

> As to their function [prediction and anticipation], Singleton (1967) suggested that they are 'models of the real world which can be manipulated to predict the consequences of certain actions and thus determine the optimum actions'. Our protocols tend to confirm that past experience is used to predict and anticipate. It should, however, not be inferred that anticipation means automatic, cue-triggered (stimulus response) behaviour. In fact, in the railway case, anticipation must never become so automatic and uncontrolled that the driver would misread a line-side signal aspect commanding him to stop or allowing him to proceed. Davis (1966) has classified the reasons given for incidents when signals are passed at danger, into 'failure to see', 'misjudgement' and 'misreading'. In each case a different form of anticipation is involved, depending on whether visual scanning, motor action or cognitive function is correctly anticipated or not. If correct anticipatory behaviour, once acquired, were rigidly automatic, such incidents could not occur. [21, 179]

Branton argued, therefore, that prediction and anticipation are not merely necessary precursors to efficient operation, but that they also *determine* the efficiency of the operation. To be able to control the vehicle, the train driver must be able to predict the outcome of his or her actions; the quality of this predictive behaviour (which may be determined in part by the quality of the information supplied) will have considerable bearing on the quality of the task performed.

> If activity and intention are determining factors in what will actually be done in using equipment, it is necessary to foresee these actualities to avoid misuse of the equipment. In this sense, anthropometric data must be applied with the function in mind. The validity of such a principle, that man's function is a determinant in use, however trivial it may appear, is difficult to prove by other than pragmatic tests. Its application to design can be vital. [15, 11]

Postural underpinnings

Having identified an important feature of the individual's behaviour which will determine the quality of performance within the system, it becomes extremely important both to understand the phenomenon and to design the system in order to accommodate it. As argued above, however, extracting all the relevant information from people is very difficult and almost impossible.

The Brantonian approach to this problem is to include consideration of the underlying physiological mechanisms that both generate and display purposivity indirectly within the framework of an operator's overall behaviour. In many cases, Branton argued, the operator's expectations from a situation can be gleaned from some kind of analysis of the body posture adopted, and so the important physiological structures are those which control posture and determine the body's structure. An analogy might be the different expressions seen on the face of someone who is engaging in difficult, dangerous or pleasurable tasks: 'Activity and intention (or purpose) will determine the operator's postures and movements' [15, 9]. The use to which postural and functional anthropometric data can be put in order to understand the operator's purposivity will be considered further in Chapter 4.

Interest, boredom and adaptation

Allied to a consideration of the purposeful nature of the operator are questions about how far the operator has an interest in the operation under control. Clearly, interest is related to purpose insofar as reduced purposivity can lead to reduced interest, and reduced interest can lead to reduced awareness of the environment.

A deeper analysis of the construct, however, leads us to consider not so much interest *per se* but the reason for the interest in the job being undertaken. Thus it has been argued that interest is expressed through stimulus-seeking behaviour. The work of investigators like Sherrington, Granit and Von Békésy, for example, shows that the nervous system adapts and habituates quite rapidly to any prolonged and unvaried stimulation, leading to cortical inhibition of stimuli and extinction of responses. Boredom, a result of reduced interest, then, can soon lead to a reduced ability to perceive and respond to important stimuli, and adaptation to the stimuli can have similar effects.

Branton recognised that this problem could arise when considering the performance of high level tasks such as train driving, particularly when the operator is *not* required to respond to a clear signal:

> However complex the environment outside of the cab, it is naturally thoroughly familiar to the driver. For most of the time on passenger trains, while going along a normal scheduled route, the signals will be in his favour and the road clear. His normal response to the sight of a signal is *abstention*

from action, so that he does not even stimulate himself by postural-muscular-kinaesthetic self activity.

Inside the cab, too, certain monotonic qualities of the environment can be found. The noises are relatively rhythmic, and so are the vibrations and oscillations. Even the climatic changes in the cab are in slow and often imperceptible steps. [11, 83]

As Branton observed, if individuals are starved of this stimulus variation within their environment they will actively seek it from elsewhere:

The evidence . . . comes from extensive behavioural records and the related intensive self-reports, often by very articulate subjects. It emerges that, at certain times and in certain monotonous conditions, moments occur when the operator spontaneously increases his activity considerably. Two reasons are given: one is an express urge to seek sensory input, a need for stimulation. If the work does not provide this input, the individual seeks or produces it himself, even if only by getting up and walking around, providing the job permits this. This interpretation of observed behaviour tallies with what is known about living organisms in general, and modern man in particular: there is an irrepressible, spontaneous need for sensory inputs. [31, 2]

Mind wanderings

As well as adaptation, a second effect of reduced interest and stimulation from the external environment is often expressed in terms of a lack of attention to the task, and to mind wandering and day-dreaming.

Even if one is not asleep, yet in a well-supported posture, whether horizontal or seated, the anti-gravity muscle action control is 'switched off'. If, in addition, the eyelids droop and visual input is suppressed or diminished, the person enters into twilight states in which the balance between attention to the environment and to the self becomes unstable.

These transitional . . . states are often frequently associated with highly original and creative thoughts and sometimes with hallucinations, in other words, with fantasies. [31, 3]

In some respects, Branton argued that such attentional wanderings could be a means by which people attempt to inject some (internal) stimulation of their own. Taking the idea a little further, he also observed that such events are often accompanied by additional physical activity by the individual:

The other reason for sudden spurts of activity is that operators report that they recall having struggled with drowsiness and realised that their 'minds had wandered' from the task. They knew that there was a time gap during which their mental activity had slipped out of control but are quite uncertain about the duration of that gap. [31, 2]

On the face of it, of course, any propensity to mind wanderings (and thus possibly missed signals) in environments where there is reduced stimulation could be perceived as a 'black mark' for the human operator's role within the system. A Brantonian View, however, takes a different approach. By considering the individual's response to the environment in terms of the moral and ethical principles which guide them in their interactions with the system, it is possible to understand how people will deal with such environments. Thus, as will be discussed later, because people do exhibit responsibility at work, the possibility of mind wanderings during reduced stimulation should lead the ergonomist to realise that the effects can lead to the operator's uncertainty and concern as to what might have happened during the period of inattention. More will be said in later chapters about the implications of this for operator stress and well-being.

Uncertainty reduction

A further aspect of the person-centred equation which must now be considered, and one which may even be thought of as having an overarching effect, is the reduction of uncertainty.

From a psychological viewpoint, Branton argued that uncertainty reduction plays a major role in everyday life; we attempt to reduce it if for no other reason than to reduce the stressful effects of unpredictability. Of course, like stimulation and arousal, individual attempts to reduce uncertainty are likely to be tempered by some concept of an optimum level. Indeed in many respects life would be very boring if all events were perfectly predictable.

Expressed in mathematical language, certainty and uncertainty concern the probabilities that an event will occur. When taken in conjunction with the 'quasi-mathematical' model of the skilled operator's behaviour described earlier, they can be seen to play a potentially important role within a person-centred approach. Thus, whereas the approach accepts that quasi-mathematical levels of information are required to aid the operator in making predictions of the likely outcome of actions, reducing uncertainty has important implications for the *kind* of information which should be provided. Such information will deal with features of the situation in which the operator enters uncertain states, rather than aspects which have more certain outcomes.

Branton's analysis of the situation places even more importance on such constructs insofar as they may have an effect on the likely stress levels of the operator. Thus, by pointing out that mind wanderings can increase the uncertainty of events which may have occurred during the period of inattention, Branton argued:

> Here are two pointers to the nature of the harmful effects of stress. The loss of TIME SENSE connected with absence of mind, and the DEGREE OF

UNCERTAINTY in the situation. Both are concepts amenable to further research. The immediate practical point to make is that if appropriate information and training can be pre-arranged and time pressure reduced, harmful effects become less likely and uncertainty more bearable. [31, 2]

At this level of analysis, then, the concept of uncertainty is clearly important when considering the nature of stressful events at work; if one is uncertain about what has happened, particularly during a period of 'mind wandering', then stress is likely to result – particularly if one also has a position of responsibility when carrying out the task (a concept which will be considered later).

Control and autonomy

Since increased control over a situation improves the likelihood that the events will occur in the desired and expected manner, it is clear that uncertainty reduction is also related to control. The evidence for 'mini-panics' resulting from 'mind wanderings' in Branton's research demonstrates that uncertainty occurs when control is relaxed. Or, put in another way, 'Only what I do is certain for me'.

Discussing the design of computer screens, for example, Branton and Shipley highlighted the importance of control as an important factor in stress causation and management. They proposed that such technology has propagated a new 'breed' of individuals – 'Houston Man':

> Perhaps his VDU screen presents an illusory picture of the true condition of any plant or process deemed to be under control? Glued to his screen, how far can he actually control the reality out there? . . . Our interest is not so much in what is actually shown to be *on the screen* but in what is behind it, in what the display is supposed to represent to the operator. The processes to be controlled are at least one step removed from the controller's direct experience. The stress arises when these displays can no longer be trusted. The very remoteness from the end product generates feelings of helplessness, a condition often reported in the literature on stress. [39, 1]

Discussing the effect of uncertainty on working practices, Branton argued that it is also likely to increase the extent of information seeking behaviour, and also the application of imagination to the job in the absence of information:

> How then does a person in a compulsively restricted situation, say, a driving cab or cockpit, respond to uncertainty? Analysis of field protocols shows that his 'adaptive response' is to create his own information. More or less consciously he takes the risk that his quasi-information may not conform to the real state of affairs. To cover that gap between his 'imagination' and the future 'real world outside', he continuously seeks for further information from the environment. The term 'adaptation' now changes its meaning from passive compliance to active endeavour. He

actively scans the surrounds with all perceptual modes at his disposal and when it is poor, say in fog, he strains his perceptual apparatus to the highest degree. [35, 502]

As a result of a poor quality of external information the operator may resort to retrieving material 'internally':

If the senses are lulled, interests and control are diminished and the ever ongoing mental activity seeks its material from fantasy by a random walk among 'internal representations'. Control over thought processes necessarily involves some monitoring of the surroundings. The more intense and exact the control over the exteroceptive senses, the greater the influence of the 'real world' on the person. [35, 502-503]

Ergonomics and organisational psychology literature is resplendent with examples of the need to provide autonomy in working situations. Its absence has been shown to be a major factor in job satisfaction, employee turnover, stress at work, and so on. By taking a person-centred view of the situation, however, the reasons for its importance become very clear: increased autonomy and personal control implies increased certainty within the working situation; we are more in charge of the consequences of our actions. This leads both to lower stress levels and to less likelihood of mind wanderings.

Responsibility

Any discussion of autonomy is incomplete without a consideration of the concept of responsibility. Autonomy and responsibility are clearly interrelated concepts; one who has responsibility ideally has conscious control over his or her own actions:

If the operator is responsible, he is a self-determined system. What he determines is his purposes and he can do this more or less at any level of present awareness. In fact, he obviously can to some extent consciously override his own bodily rhythms – provided his purposes are strong enough. In this, his determinant is the consciousness of the consequences of his own actions before they are carried out. That is the essence of his rationality, implied by investing him with responsibility. [31, 3]

As became evident in Chapter 1, the Brantonian View places considerable emphasis on the concept of the *responsible* operator. Thus, people at work generally act out their duties in sane and rational ways; in that respect they behave in a reliable way. Having a sense of responsibility, however, can also have negative effects – particularly in the likelihood of increased stress resulting from it. This becomes particularly important in situations when mind wanderings may occur – the resultant mini-panics can be considered to be a direct result of the responsibility which the operator feels over the situation:

Unfortunately, individuals work within systems in which technology demands more than is within their natural capabilities, over which they, anyhow, have less than full control. The train driver, unlike the car driver on a minor road, cannot suddenly decide to stop and take a rest when he notices that he is tired. He is responsible for the train and may overrule his bodily functions up to a point. Most often he does not even recognise the point at which he becomes overloaded until it is too late. [40, 14]

Summary

The person-centred approach to ergonomics, particulary the Brantonian View of it, has opened a number of intriguing possibilities both for further consideration and study. In particular, the approach highlights the personal philosophies underlying physiological processes involved when people do their jobs – particularly those which may play a part in lessening the uncertainties to be contended with when trying to carry out responsible actions.

Specifically, the person-centred approach views the human operator as:

- purposive
- information seeking
- uncertainty reducing, and
- responsible.

If such factors are not accommodated within the system, however, the result is likely to be to create an environment in which people are unable to express their essential human-ity and person-ality. Systems which lack this approach will thus tend towards being mechanistic creations which concentrate on the outcomes of the operation rather than on the quality of the process.

Chapter 3

Unselfconscious behaviour

Chapter 2 illustrated some of the fundamental processes which a person can bring to a situation and which helps to shape his or her behaviour when interacting with it. People do not simply passively interact with the environment, merely responding to ergonomically designed displays using ergonomically designed controls within an ergonomically designed environment. Rather, the importance of concepts such as purposivity, anticipation, interest and boredom serve to highlight the person's central role in the effective fulfilment of the task.

When interacting with a system, people are active and creative doers rather than passive recipients of information. They develop mental models of the system: how it operates, how it should be operated, how to interact with it, how they should interact with it. Thus, an understanding of both the models themselves and how they develop is crucial to an understanding of how the system can best be designed to accommodate and even enhance the models and, from them, the interaction.

This person-centred extension to the traditional ergonomics philosophy and practice was made more pertinent by much of Paul Branton's work and thoughts. Indeed, he considered a central necessity to be the study and understanding of the roles of various 'unconscious' processes in determining and developing these mental models. As will be seen in the next chapter, he used his realisation of the existence of such processes to develop interesting measuring tools to investigate a variety of phenomena which impinge on the operator's efficient capacities: stress, discomfort, performance rhythms, guidance control, and so on.

Within both experimental psychology and ergonomics, however, the study of conscious and unconscious behaviour has traditionally met with some degree of disapproval. Rigid behaviourist principles allow only the study of phenomena which were strictly observable and of which repeated measures could be made. A process which is being carried out without full consciousness cannot be observed, it was maintained; ergo, it cannot be studied. Such doubtful propositions can still be seen within the modern ergonomic literature. Branton, however, argued that to disregard such processes is to dismiss crucial features of the operator and the ways in which he or she interacts with the system:

> I shall draw on work done with train drivers to show just how much 'non-conceptual mental work' must be involved in skilful pursuit of purpose.

27

Consciousness is something which some of our colleagues in that most recent branch, cognitive ergonomics, seem to by-pass. They concern themselves mainly with human functions that are verbalisable and computational, overlooking vast areas of non-linguistic, non-conceptual, mental activity. Perhaps they hope the problem will solve itself later. However, as I pointed out elsewhere [28], in the process large sectors of real life are ignored, while the problems of explaining such basic functions as selective attention of directed awareness still remain with us. The largest part of our knowledge is still 'tacit knowledge' (Polyani, 1958), a concept which has an important role in the general case for the purposive view. [40, 6]

Rather than merely criticising those who argue that it is not possible to study the degree of unconsciousness in behaviour, however, Branton also went on to propose ways of investigating such phenomena. He emphasised that in many cases it *is* possible to study unconscious processes, by observing their effects on the operator. Discussing the importance of postural changes, for example, he observed:

Postures are rarely contrived, hardly ever deliberately taken up and wholly beyond one's awareness, except when one's attention is drawn to them after the event. Nobody says, 'Now I am going to cross my ankles', or 'Now I will put my hand on the armrest.' (Yet no one outside the person can tell whether a posture was taken up deliberately or not.) The records definitely point to the presence of processes, not directly observable, perhaps not observable at all by the instruments, but which must necessarily be going on out of sight *to produce that which is indubitably observed.* The records represent the circumstantial evidence. [40, 5]

Again, the implications of this kind of approach will be considered in more detail in the next chapter. For the present it is important to realise that the study of unconscious processes is not just a desirable activity for ergonomists to undertake; it is a necessary activity because such processes, or at least their intent, can have major practical implications for the design of the system. Indeed, Branton sometimes referred to them as 'un*self*conscious' behaviours, a term which implies that the 'unconscious' is only relative, that unconscious processes may have some conscious components. It is rather that the 'self' has no immediate or explicitly verbalisable awareness of them.

The importance of the self

This discussion of the role of unconscious behaviour within a person-centred approach has led to an important question: If the self is important in this conceptualisation of ergonomics, what is the nature of the self? And, equally, granted the existence of the self, which of its aspects are important for understanding the nature of our interaction with our environment?

Philosophical discussions surrounding the controversial concept of the self have a long history, extending back to Aristotle. In relation to behaviour, one of the first among more modern thinkers to attempt an analysis of the concept was the psychologist William James who felt that our sense of self rests on the amalgamation of two components. First are aspects which we know about ourselves – our self concepts (the 'me' in James's terms). Second are the features of our psychological make-up which enable us to process and make sense of this knowledge (the 'I'). In some sense, then, the difference between the 'me' and the 'I' is one of direction. The 'me' represents our self perception, how we feel about ourselves and our actions. The 'I' is more of our public persona, how we use the 'me' to interact with the world. James felt that the 'me' and the 'I' combine to form the self.

Whereas many of one's activities which pertain to the self are bound to fall within conscious control (self awareness), Branton felt that the unconscious features of the self – the un*self*conscious aspects of behaviour – are equally important aspects of our behaviour and ones which should be understood. It is through them that we might better comprehend some of the processes which underlie the individual's behaviour.

Whereas a complete exposition of the nature of the self, its structure and components, is outside the realms of this introduction to person-centred ergonomics, to understand fully Branton's view of the discipline it is important to discuss some aspects of this concept. In particular the link between the self and subsequent behaviour should be considered. In establishing such a link Branton suggested:

> My own line of argument would go on like this: We begin by accepting Kant's transcendental argument that humans cannot know the 'real truth' – whatever it may be – with *absolute* certainty. For the possibility of 'knowing anything at all', we must therefore ultimately resort to sources of knowledge which lie *within* ourselves. Such knowledge as we have, we possess only as actually thinking beings who process by their own effort whatever information they receive from outside. From this we may validly conclude no more than that the most basic human faculty is the ability to think. [28, 4]

This statement, of course, returns to Descartes's famous proposition: 'I think, therefore I am'. In making it, however, Branton was not simply endorsing narrow Cartesian logic. He argued strongly that our ability to interact with the environment and, as ergonomists, our ability to design systems which facilitate this interaction, rests largely with our ability to *use* the knowlege which is 'within ourselves'. The nature of this knowledge, therefore, becomes increasingly important.

In relation to the current discussion, knowledge could be said to be derived from two sources which almost parallel James's 'me' and 'I'. The first, external knowledge, appears from sources which are external to the individual, from others, from the system with which he or she is interacting, from past experience, and so on. Internal knowledge, on the other hand, derives from the 'me' part of the self. This includes both conscious and unconscious information about our physiological states, knowledge of our actions, purposes, and so on. As individuals we have to be able to differentiate between these two, often competing, sets of information.

As an analogy, Branton invoked the developmental processes which occur:

> To produce the known results, the self in all its manifold manifestations must be represented internally. We can see how it works if we combine observation with indirect knowledge. We observe an infant lying in its cot, repeatedly touching first the side of the cot and then its own toes. Even with our limited knowledge of neurophysiology, we know that there must necessarily be a difference between perception of 'not-self' and 'self'. In the former case, one set of sensory signals is received in the infant's brain: from the fingers only. In the latter case, there are two simultaneous sets: from its fingers and from its toes. The simplest generalisation of this experience is: one sensation = not self, two sensations = self. [30, 7]

The introduction of some kind of internal knowledge into the discussion is particularly important in relation to our knowledge of the effects of bodily rhythms on performance. Thus, through internally generated information the responsible person *knows* of their presence (although they may not be explicitly and consciously aware of them) because they are self-generated and are thus certain. To overcome the difficulty of explaining how we can, at the same time, be both aware and unaware of something, Branton argued that much of this could be related to the concept of 'pre-cognition':

> A brief critique of cognition usually relates perception to conscious knowing, thinking and understanding, to intelligence and the intellect, to linguistic expression, to the use of concepts and to 'handling' symbols for problem solving and the like. Philosophers, too, use the term in their theories of cognition. But, as Bruner (1980) shows, and many other psychologists frequently remind us, we know more than we experience and we experience more than we perceive. There must be more mental activity besides *conscious* cognition and we must have more knowledge beyond that of which we are conscious at a given time. Supposing some unconscious knowledge is immediate knowledge, might it not be that such knowledge is originally non-cognitive? Might it not be non-conceptual and non-linguistic? [28, 5]

The importance of such an approach then becomes clear. Unconscious (unselfconscious) behaviour helps to shape our conscious behaviour through the development of pre-conceptual thought. It thus helps to shape the way in which we approach an activity and how we eventually carry it through:

The existence of pre-conceptual thought has long been postulated in various guises and, in my view, there must be a stage in thinking about something before its proper concept is formed. Indeed I would go further and postulate explicitly the existence of some kind of thinking which need never rise to the clarity and consciousness of a specific concept. [28, 5]

Thinking and doing

To argue for the importance of unconscious and 'pre-conceptual' behaviour to eventual conscious activity, however, we must also consider the bridge between this kind of thinking and eventual 'doing'. What is its nature? How does it arise? And how can we take cognizance of it?

Branton offered a solution to the problem by arguing that the very nature of the thought process is one which raises the unconscious to the level of conscious behaviour:

> In the end the sustained raising of any unconscious knowledge must be an *inner* process. The person himself might do it by exercising critique on his own non-conceptual knowledge when he has an intimation of some item of knowledge-before-conscious-experience and then inquires into his pre-assumptions that made this possession possible. The process of bringing immediate knowledge into awareness therefore presupposes the autonomous person as a thinking being. [28, 6]

The bridge between these thinking and doing activities, Branton argued, is created by the individual's propositions about the information which is being assimilated and manipulated. Such propositions, or quasi-propositions because they are not really verbal and often not even conscious, contain immediate and certain knowledge. I know where my foot is without having to say to myself 'There is my foot', or even without having to think about the foot at all. Furthermore, perhaps more importantly, it is the way in which such material is dealt with which forms the structure of the bridge – the individual's own internal model of the system's structure:

> Before a bridge can be built from knowing to doing, we must go into the role the propositions play and how they allow the person to take in information from the world around him, especially the social world. If the only pure proposition states *self*-consciousness, that person might be autonomous but would also be egocentric. Other implicit statements must therefore let the outer world in. Their role is to *relate* or *separate* events or features and attributes of objects of the world *in our thoughts* . . . These propositions to the self are implicit and need not be in natural language form [28, 6]

By helping to build the bridge between thought and action, therefore, from the ergonomist's perspective such quasi-propositions play an important role in determining how the individual will act within his or her environment.

To expand the concept further, Branton argued that two further basic and complementary mental faculties need to be introduced: conjunctions and abstractions. These concern how we relate quasi-propositions to each other to build the bridge's structure (the model of the system), and the ways in which we extract and combine the information from our internal and external worlds in order to create them.

Propositional relationships

To explain the nature and development of the quasi-propositional relationships, Branton invoked a concept which is similar to the basic view of thinking as associations of ideas and concepts. In addition, he followed Kant's arguments that causal relationships between concepts also require the presence of some kind of 'meta' associationist structure:

> Hume proposed laws of association by repeated proximity of two external events to explain the sequential nature of mental activity. Not only would the conjunction of events be recalled, but we would also come to expect the conjunction to have the same effect each time. The assertion of causal lawfulness based on experience alone provoked Kant to expound the need for postulating the existence of *a priori* synthetic propositions. [28, 7]

This kind of argument is similar to the associationist's views of hierarchies of associations which lead to the formation of solutions in the thinking and problem solving process.

The process of abstraction

The bridge between thinking and doing, and the development of quasi-propositions to serve it, relies also on how we abstract the two different kinds of information within the self (the 'me' and the 'I') and draw out the salient features. The process of abstraction is thus one which is central to understanding the role and nature of the unconscious processes. To comprehend this process, Branton pointed to two levels of analysis: the physiological and the cognitive. At a physiological level, Branton pointed out that neuropsychological studies have demonstrated feature-analysing systems within the brain which help to develop this abstraction process:

> The difficulty of describing the psychological process of abstraction has been much reduced lately by advances in psychophysiology, when it was shown experimentally that during a very early stage in perception an object is analysed and each of its features or attributes 'abstracted' for internal representation in a separate cell or group of cells. For example, the specific wave frequency band or the colour 'attributed' to an object activates one or

a group of cells, the straightness of outline of its shape another group, the obliqueness of that line yet another, and so on. It is therefore possible to say that, *in our heads*, the sum of the predicates of all the propositions we implicitly make upon perceiving an object IS this object for *us*. [28, 7]

The implication of the final sentence in this quotation should not be missed. Thus, by arguing that the quasi-propositions constitute the internal representations of the object, and that the ways in which we abstract the information determine the 'structure' of the object, Branton also maintained that the individual plays a central role in determining the final response to the environment. Thus, the act of turning a control, for example, is not defined by the system but is derived from the individual's quasi-propositions and abstraction of how the system operates. Thus '[the information] impinges upon an organism already structured and prepared to categorise environmental features into attributes and abilities' [20, 5].

In this way, then, Branton moved from a psychophysiological view of the abstraction process to a cognitive one. Thus the actual process of abstraction implies an ability on behalf of the operator to categorise the quasi-propositions and to discriminate between them. He does this by the simple assertion that there must be some 'internal' cognitive structures which help to condense the wide variety of possible quasi-propositions to 'manageable' proportions.

It would seem that one consequence of asserting this [that the experience impinges on a pre-prepared individual] is that the number of attribute-categories provided by the system must be sufficient to discriminate between the variety of possible stimuli. Is this true? Must there be as many pigeon-holes as there are attributes likely to be encountered if the manifold of natural events is to be explained? [20, 5]

His response to his own question is that this is not the case:

The finding that predicates are abstracted from their objects has profound consequences for understanding the mind. It helps to explain how a vast collection of things can be internalised, stored with economy and reassembled again in recall. Things may thus share attributes with other things and their representation need not be concentrated in one point. [28, 8]

Abstracting the abstract

Although the meaning of abstraction has so far been discussed in terms of reducing information down towards manageable proportions and extracting from it salient features, the word 'abstract' has another meaning. It is one which also has implications for the person-centred approach to ergonomics understanding.

An abstract object is one which has no form – a mental image, as it were. The process of abstraction, therefore, is also related to the process of imagery, a point which Branton emphasised when he argued that the individual's concepts of the system are 'precepts idealised in the process of internal representation' [20, 1]. These internal representations, then, are abstract ideas which must be transformed into concrete actions. Discussing the nature of the train driving skill, for example, Branton argued:

> That such internal representations and processes exist is not in doubt. Where could they be? Surely not at one single point in the head? In what form could they be held? 'Engrams' and 'structural cell assemblies' would be far too rigid. . .The question then turned into: 'What exactly is represented in the mind of the driver?' Some call it memory, others experience, and pass on. I had to conclude that, to act as he did, he must *possess knowledge* of the whole of his 'railway world' in all its complexity, together with the immediate consequences of his own actions, as well as all the postures and movements required for success. He just could not have achieved his skilled action without this 'intelligent knowledge base'. [40, 9–10]

> . . .The mere fact of our having in our head a kind of ready-to-use replica of the outside world makes us relatively independent of our immediate perceptions of the environment and thereby autonomous. By this means we can break the natural chain of causal determination and can become self-determinate. We are then no longer merely reactive, but self-activated. [40, 10]

The concrete expression of this abstract representation, Branton argued, underlines the possibilities for being able to study such unconscious behaviour – and thus use it to facilitate the ergonomic design of the system. More about this line of argument will be considered in Chapter 4. Here, though, it is appropriate to conclude with Branton's observation that such representations are often expressed in conscious behaviour:

> It is a common observation that very articulate people, even eminent scientists, sometimes underline their speech with gestures, 'thinking with their hands', apparently without being aware that they do this. Being now on the look-out for the unknown, we get here, again, a glimpse of submerged parts of the iceberg, representations that were initially unconscious but become public by behaviour. [40, 11]

The self and activity

The foregoing discussion has introduced the nature of the self and has discussed aspects of its role in determining action. It has also considered how this conceptualisation may lead to practical applications within ergonomics and measurement. The discussion must now shift towards considering the nature of the activity embodied within the self concept: the concept of the active self and its preparedness to act.

That the self is active and purposeful is by now a tenable proposition; that this activity could be better encouraged and facilitated within the ergonomics of designs is also now a compelling view. What has not yet been explained are the reasons for the activity. Branton argued that answers to this kind of question lie in the importance of the information which we receive and on which we first unconsciously and subsequently (perhaps consciously) act; the internally generated information from such sources as our muscles play a particularly important role in this respect. Related to the concept of the self, this concerns the 'me' side of the equation.

Taking posture, for example, Branton pointed out that the body is so designed that it continuously has to 'fight' against the effects of gravity. Thus it is constantly active and this activity supplies the information upon which the self and its representations are composed:

> . . .everyone knows that people do not usually sleep standing up or sitting upright, and that they 'nod off' when seated comfortably. The mystery was that subjects tend to relax more easily (at certain times of the day) the further the head is behind a vertical above the hip joint. . .Backrest angles of 30 degrees or more from the vertical seem to be 'drowse-inducing'. If reduced arousal and sleep are inversely related to the angle of posture, what is it about uprightness that keeps one aroused? Conversely, what is absent in sleep that is present in the waking state? The answer to the second question is far from trivial and its full import took a long time to become clear. It is, of course, the continuous fight against gravity which pervades the whole body and its psychophysiological control system. . .Could it be that the undoubted dynamic instability of the body in standing and sitting directly affects alertness and consciousness? [40, 4]

In a previous paper discussed above, the same evidence was used by him to argue a slightly different, but equally important point: that the effect of reduced self-activity will ultimately lead to a loss of purposefulness and increased likelihood of mind wandering and day dreaming. Thus, the proposition is advanced that the internal stimulation which is derived by the active self abets two further processes: first, an increase in the seeking of external stimulation, and second, the reduction in a tendency to become more 'abstract' in thought:

> . . .both perceptual intake and body movement is suppressed during sleep. But even if one is not asleep, yet in a well-supported posture,. . .the anti-gravity muscle action control is 'switched off'.

> . . .a connection is thus suggested between secure suspension of the body and reduced information intake on the one hand and enhanced abstract mental activity on the other. Conversely, it is arguable that two factors which normally prevent one from being conscious of those states of diminished external awareness. . .are the active search for stimulation, particularly visual, and anti-gravity muscular work. [31, 3]

Preparedness and purpose

The discussion of the nature of unselfconsciousness and of the self leads us
finally to consider some of the practical implications of such concepts and ways
in which they can be incorporated into ergonomic design. As far as Branton was
concerned, this path lay firmly in realising that the role of the self and of the
state-preparing unconscious behaviour was intimately bound up with preparing
the individual to behave in certain ways.

Continuing with the tradition of the Critical Philosophy, Shipley and Leal
(1991) propose that the active self provides for the individual a source of
certainty of which he or she may or may not be consciously aware; it is so much
taken for granted. This certainty, or reduced uncertainty, enables individuals to
cope better in, and to deal more appropriately with, the world. Clearly, this is an
aspect of which ergonomists need to take cognizance, and was discussed in
Chapter 2.

As well as interfering potentially with the successful outcome of a task,
through interrupting the quality of the pre-planning process, increased
uncertainty may also lead, Branton suggested, to stress and overload. This can
be alleviated: '. . .if appropriate information and training can be pre-arranged
and time pressure reduced, harmful effects become less likely and uncertainty
more bearable.' [31, 2].

In addition to concepts of certainty and uncertainty, Branton has argued that
the active self is purposeful and purpose represents for the individual the likely
future state of his or her actions. Thus, we are in some measure able to predict
the outcome of our actions and perform acts in the way which will most likely
achieve the desired results:

> Of all the considerations involved in describing this clearly complex skill
> [train driving], the most far-reaching one turned on the nature of internal
> representations and mental operations. When a highly responsible operator
> performs a task so accurately and safely over time and distance, and mostly
> without direct or immediate sensory feedback, some very essential data
> must be represented internally. Simple visual memories in the form of static
> targets are totally insufficient to explain the accuracy with which train
> drivers bring the trains to a halt just before signals and a yard or two before
> the buffers at the terminus. [40, 8]

Thus, for Branton the active self clearly has purpose and, by implication, it is
incumbent on ergonomists to design systems which enhance this self-purpose
interaction.

The question can now be posed as to just what these purposes are. If purpose
is the internal representation of some future state, at what time in the future is
the individual aiming? Branton's answer to the question was to suggest that
individuals need to be able to predict the effects of their actions, and so the
'future' in this case will be 'a-little-after-the-future'. In other words, when
displaying information about the likely outcome of an action, for example, the

system needs to indicate what is likely to happen just *after* the outcome, rather than 'simply' the outcome itself. This prediction, of course, is crucial for the peace of mind of those who carry responsibility.

Summary

This chapter has considered some of the more philosophical foundations of Paul Branton's approach. By viewing the behaviour of an operator at work as having its foundations in the individual active 'self', in the concept of one's '*self*', Branton emphasised that ergonomists should take a major step back from the system and look at it from the standpoint of the individual's understanding, wishes and purposes. Thus the active, thinking, doing operator brings to the system a set of ideas (quasi-propositions, models and abstractions) which are likely to determine how he or she will effect the interaction.

Chapter 4
Measuring behaviour

Whether operating in a traditional manner or within a developing person-centred framework, ergonomics is very much a science – a science of human interactions with the environment. As such, it progresses not so much by armchair philosophy but empirically by the collection and interpretation of data. The person-centred ergonomist, therefore, must develop data-gathering techniques which emanate from the central philosophies of the discipline, and follow the canons of scientific rigour in the fullest sense.

In expanding the ideas and concepts that were central for him (and which have been discussed in previous chapters) Paul Branton was able to develop ways of gathering data which would maximise the role of the human within the system. The techniques which he used are grounded in the basic methods employed by most psychologists and ergonomists. However, they were guided and informed by his developing philosophy which influenced the *ways* in which such traditional methods as behavioural observation, psychophysiological measurement, protocol analysis, and interviewing procedures were to be used, and their data interpreted.

For him the central importance of research and systematic method within ergonomics places it above many other less theoretically and scientifically inclined technologies. He believed that the development of the discipline lies firmly in the production of ideas which will help to build new methods of data collection and interpretation. Discussing the beneficial aspects of an ergonomics perspective, he suggests:

> My third praise is for the practical significance of ergonomics. By practical, I mean something which helps to answer questions of what should be done. The lay person, told that ergonomists study how to avoid 'human errors', immediately understands the social usefulness. Other disciplines may serve as technologies only, i.e. as mere adjuncts to theoretical science. If anything, Ergonomics as a technology is explicitly in the business of fitting technology to its users. Its objectives, its purpose, is man. As a research model in its own right, Ergonomics deals with hypothetical imperatives of the 'if. . .then' kind, yet its practitioners can never altogether escape from facing questions like, 'What is it all for?' 'What are the values of the system?' [40, 2]

A person-centred approach to measurement

Epistemology

Branton argued strongly for an epistemological approach to research:

> To find ways out of this conceptual maze, one guiding thread may be offered to the researcher. It is to adopt an 'epistemic' strategy: to ask himself in the first place at each stage, 'How do I know whether this or that statement by the operator is true?' 'What is the source of my knowledge?' 'How direct or indirect is my perception of the measurement?' 'How far is my conjecture based on an analogy and how far does it penetrate to the 'real thing'? Having thus become conscious of their own inevitable bias, observers (and their readers) are better able to speculate on their subjects' knowledge, values and actual purposes. [35, 505]

And in the following extract which discusses the analysis of train driving skills:

> The assumption of purposivity, or goal-directedness of skill, combined with critical analysis of the operator's stored knowledge, leads to what may be called an 'epistemic' approach. We watch successfully completed, skilled actions, then dissect our observations and ask at each stage what operations are needed to explain success. 'What would the driver *need to know* to succeed? What *did* he know from the outset? What could he *not know* (without scanning the environment)?' And then we can ask him more about what he thought he knew. By this means we widen the scope of study to include behavioural observations and need not be ashamed to speculate and interpret behaviour purposively. If in the process this leads to potentially quantifiable information, so much the better. [22, 157–157]

The penetration of the Critical Philosophical method is evident in these arguments.

Multiple approaches

Such processes are conducted within the traditional ergonomics framework of obtaining measures from a wide variety of inputs to the problem. Extending the above argument, for example, Branton continued: 'Even though it is necessary to trust in self-reports elicited post hoc that will not be enough; the endeavour must be to obtain convergent information from a number of disciplines and angles'.

In his observational study of train seats [5], for example, Branton demonstrated that no single measure which he recorded and which might help to derive some 'index' of seat comfort was adequate. Observation of body movements over time, duration of postures, frequency of postural type, and so on, yielded interesting information but each, on their own, did not provide any conclusive information which was sufficient to help predict which of two seats would be the most 'comfortable'. It was the amalgamation of data from all these areas, and others, which aided the solution:

> To extract from the survey and the films information meaningful to seat designers is not as straightforward as it may seem.
>
> Mere observational data of the use made of seat features do not necessarily provide useful answers. Taking for example the findings on the use of headrests: one set of observations does not allow us to decide whether people do not *want* to use the headrest or *cannot* use it. We have argued that there appeared to be considerably more need for head support than the use of headrests showed because a further 17 per cent of sitters contrived to rest their heads on their hands.
>
> Another instance, regarding backrest use, was that, contrary to the impression given by most sets of data showing Type II [seats] to be better than I, the tall slumped as much in II as in I. A close look at the films showed that even in slumped postures their shoulders touched or almost touched the side wings, whereas the short sat well back without touching the wings. It was the shoulder region which was critical. Because the wings got in the way of their shoulders, we concluded that the tall *could not* use the backrest fully and therefore slumped. [5, 49]

Thus, it is apparent that the Brantonian approach to measurement has two prongs: first to extract information which might lead to a solution of the problem by using a variety of measures – and not to rely entirely on one kind. Second, and possibly more importantly, in this process of extraction the approach is also to develop the kinds of questions which will lead to a more person-centred analysis of the situation: to link seat headrest use to other ongoing individual activities which will provide head support, for example, and to consider backrest support in terms of the operator's abilities to use the backrest. Expressed in this way, many of the philosophical principles expounded in the previous chapters come to the surface: principles associated with the user's purposes, activity, unselfconscious behaviour, and so on.

The importance of empathy

The Brantonian approach relies heavily on the researcher being able to 'empathise' with the subject. Indeed, Branton argued that unless the ergonomist *can* empathise with the individual's situation she or he cannot fully enquire into, or understand, its effects on the individual. It is rather like trying to imagine oneself in the other person's shoes:

> As an observer, one is greatly helped then if one has a capacity for empathy with the operator. That does not mean that one should be biased or otherwise favour the particular person observed, but that one should make

a deliberate effort to enter the situation through the eyes and mind of the observed. [40, 12]

As an example of the direction in which this kind of empathetic approach may lead the researcher, Branton and Shipley [39] discuss the kind of work involved in monitoring a VDU, and how it can lead to stress. Whilst watching the BBC TV News film of the reactions of 'controllers' at Houston, they empathised being the controller at the instance of the Challenger space shuttle disaster. They went on to discuss the controller's work in terms somewhat different from the traditional ergonomic terms of information monitoring and processing:

> Our interest is not so much in what is actually shown *on the screen* but in what is behind it, in what the display is supposed to represent to the operator. The processes to be controlled are at least one step removed from the controller's direct experience. The stress arises when these displays can no longer be trusted. The very remoteness from the end product generates feelings of helplessness, a condition often reported in the literature on stress. [39, 1]

Branton and Shipley went on to consider the extent to which the controller works in social isolation, and how far she or he relates to the information being provided by the machine. Empathy is not always confined to feelings about people, especially in our modern technological world. Controllers, too, have empathy, but this may lie more with their machines as powerful extentions of them*selves*:

> Our assumptions about controller-operators as purposive persons are that they are at their best when interacting with a machine as extensions of themselves (this requires a good new theory of models-of-self). . . [39, 2]

Empathy, on the researcher's part, has important implications for decisions to be made about the ways in which studies are carried out, which variables are to be considered, and how interpretations are to be made of the data. Discussing his comprehensive study of train drivers' skills, for example, Branton argued that:

> For our study to be of any practical use, we had to try to catch at least a glimpse of what was being stored in memory and in what form, i.e. what inferences the man made at any given time about the states of the rest of the system; only on the basis of such information might it be possible to design suitable aids and direct training to enhance the driver's performance. Thus the first methodologically important decision was to *seek a purposive or prospective type of explanation* (Taylor, 1964) of driver's behaviour, rather than regressing from an observed effect to 'a cause'. [23, 156]

This quotation illustrates Branton's view of the connection between empathy, purposivity and epistemics. Only when ergonomists try to see the situation as it presents itself to the operator (who is pursuing a purpose) will we be able to understand the skill which is being employed. Without such insight ergonomists would simply be attempting to understand the external forces which made the individual act as he or she does – rather than the internal knowledge which the person brings to the situation.

The need for definition

When considering the 'machine' or equipment side of the system, the definition of terms is relatively straightforward. A desk, for example, has specific parameters which can be defined and measured using standard techniques: its height, depth, weight, composition, and so on. Although researchers may argue about precise nuances within definitions, and the relative importance of different parameters, everyone knows what 'height' means, will be able to measure it, and will have some common understanding of the implications of the height measurements.

The case is not so clear for the human side of the equation, however, and the problem becomes even more apparent when a person-centred approach to ergonomics is taken. Many of the important components within such an approach hold different meanings for different people and, since the approach requires an understanding of the operator's viewpoint, the importance of definitions becomes even stronger. A person-centred view of the desk height, for example, would consider the user's functions, the way in which the user has 'control' over activities involving the desk, use of the desk to reflect desired activities, the nature of the user's pre-conceptions of the desk and the way it should be used, and so on. Answers to the kinds of questions generated by such an approach are not so simple; they will depend on the 'definitions' offered for relevant aspects of user behaviour: function, control, wishes, intentions, and the like.

A particular example of the importance of this question arose with Branton's investigations of train passenger comfort. Comfort is a subjective response to the environment which, traditionally, is measured using some kind of scale on which the subject 'rates' her or his feelings. Scales of this kind are generally bi-polar: comfortable-uncomfortable pleasant-unpleasant.

Branton argued, however, that such scales have been evolved from essentially 'static' concepts of comfort; that they imply that comfort is a unidimensional state of invididual experience. Drawing on the work of Teichener (1967), he argued that there is no fixed or ideal set of environmental parameters which will induce in the individual a state of 'comfort'. Rather, the parameters induce physiological states which are continually changing – rather like our homoeostatic response to the thermal environment which helps to maintain an optimal thermal balance within the body. Viewed in this way, comfort is a varying quality which reflects the body's response to these continually changing states.

We shall return later to the two ideas embodied in this argument (the importance of the changing states within the body and of a homoeostatic system). For the present, it is important to understand how such thinking led Branton to consider the concept of comfort and thus the nature of its measurement.

Considering comfort from the viewpoint of the individual, rather than from the perspective of the equipment designer, Branton argued that it is impossible to measure the state since, being a dynamic complex of changing varying physiological states, it does not exist as such. In the same way as it is impossible to define health without introducing some concept of ill-health. How do you know you are healthy, if not because you have no symptoms of ill-health? So it is impossible to define comfort – other than in terms of varying degrees of discomfort:

> This is because the absence of discomfort does not mean the presence of a positive feeling but merely the presence of no feeling at all. There appears to be no continuum of feelings, from maximum pleasure to maximum pain, along which any momentary state of feelings might be placed, but there appears to be a continuum from the point of indifference, or absence of discomfort, to another point of intolerance or unbearable pain. [13, 65]

Branton then continued to argue that, defined in this way, comfort can only be reasonably measured in terms of general awareness 'as part of the general problem of selective attention to bodily and mental functions. . .the person is not just passively sitting there, waiting to be stimulated.' [13, 65]

If it is to be measured using scales at all, therefore, comfort must be considered as an attribute arising from a dimension in which it plays no part or, if it does, then it is in terms of its opposite: discomfort. Thus the comfort dimension is actually expressed by a scale of 'no discomfort' to 'maximum discomfort'.

The servo-based human

So far, the argument has emerged that the individual at work is an active being rather than a passive reactor to externally defined events. He or she:

- has motives, purposes, interests and responsibilities;
- takes cognizance of both external and internally generated stimuli in order to decide how to respond to the environment; and,
- responds to the presence of continually fluctuating bodily states.

It is now appropriate to consider these fluctuating systems in more detail. Within the development of his philosophy, Branton worked in three distinctive areas: posture, error production, and alertness. In each, he illustrated the importance of considering the varying nature of these inputs and of such fluctuations in the measurement of a variety of challenging phenomena such as comfort, stress, and skill.

Understanding the fluctuations on their own, however, is not enough. The

Brantonian approach, it should be stressed, is to consider the purposeful inputs which the individual brings to the system. These inputs, Branton argued will serve to alter the underlying fluctuations and may sometimes be susceptible to measurement. Discussing the physiological 'pathways' whereby such inputs may arise, for example, Branton has argued that:

> The constant stream of stimuli from outside the body, competing with what goes on inside in the business of keeping alive, upright and active, is too much for us and incoming information is severely filtered. . .

> Only the perceptible and unusually strong changes in the incoming stream are reported 'up the line', though not necessarily brought to clear awareness. However, it is the organism which 'decides' what is unusual and therefore interesting. The decision mechanism would seem to operate on the purposive levels and is capable of overriding other levels of function. Purposive or programmed activity, even if removed from awareness, will set new limits to the self-stabilising system. . .

> As long as the environment remains safe and adaptation enables him to ignore it, the person's purpose will continue to compete with bodily demands for change and variation. The organism will seek stimulation and, if no other change is possible, a self-induced increase in sensory input will occur to compensate for otherwise monotonous environmental conditions. The person. . .can thus be regarded as a self-perturbing, rather than adaptive, set of systems. It is this author's conviction that the self-perturbation is amenable to observation as it is likely to manifest itself on the overt behavioural level of function. [13, 66]

By considering such fluctuations at an empirical level, Branton argued, it is possible to measure the individual's inputs to the underlying events: posture, error corrections, and fluctuating awareness.

Posture

Maintaining a posture, such as during data entry or even just sitting, requires activity, which is often unconscious. As Branton argued:

> . . .once you are seated. . .the problem of holding yourself upright and otherwise maintaining your position deserves much more attention than it receives. . .For, what has dropped in the seat is not a bag of bones, fat and gristle – it is a living being that will not sit still. In other words I greatly regret the apparent unconcern in most studies that sitting is an activity. [19, 4]

The above assertion, however, begs an important question: 'what is the activity engaged in when maintaining a posture?' The answer to this, Branton suggested, lies in the fluctuating nature of the forcing system which is liable to lead to instability unless the body *is* active: the body under the effects of gravity.

The upright or sitting body is an inherently unstable system. Standing upright, for example, the centre of gravity of the body passes just in front of it. We thus need to be equipped with efficient servo-mechanisms to enable us to maintain an upright posture in the face of such adversity; this is performed by the balancing mechanisms in the inner ear, aided by feedback from positional receptors in the muscles – particularly in the neck and optic regions.

The postural activity, therefore, is one of fluctuating body states in order to maintain the desired posture:

> . . .posture maintenance and control are fully dynamic processes, going on all the time – a two-way traffic across the borders between body and mind. Simple physics tells us that body parts arranged in the seated form must collapse – unless they are held in place by muscles controlled by a conscious mind. Try to lift your feet an inch off the floor as you sit and you will begin to see what I mean by on-going maintenance activity. [19, 1]

Aids to the maintenance of posture, and to the prevention of the onset of muscular fatigue, therefore, should be so designed as to ensure that these fluctuating states are facilitated, rather than blocked.

As an example of how this can be considered, Branton proposed a homoeostatic theory of seating comfort which has similar elements to that discussed earlier. In this system, comfort is conceived of as being the optimal state between two conflicting body states or requirements: stability and pressure reduction. By measuring such states, and their conflict, Branton argued, it is possible to have some measure of comfort. When seated in a chair the individual will gradually experience the build-up of pressure in the lower thighs, due to compression fatigue, and of muscular fatigue caused by the need to maintain a particular static posture. Slight shifts in posture will help to alleviate such sensations, rather in the way that slight changes in the direction of blood flow or the secretion of the sweat glands help to alleviate the build-up or loss of heat in the body. Such a homoeostatic model, therefore, leads to the possibility of measuring the 'discomfort' of seats, by observing postural changes – or fidgets.

Such postural changes are generally very slight and are not always amenable to direct observation. To overcome this problem, Branton used speeded-up film and video-tape recordings; by playing the recordings back at a much faster rate than that at which they were recorded, it was possible to speed up slow and otherwise imperceptible behaviours so they became perceptible. This technique was used extensively in his evaluation of train seats:

> . . .individual frames of observations passed the observer's eye 160 times faster than they were taken. The compression of time (5 hours into 2 minutes) brings out drastically the dynamics of the sitting situation. The film of one subject is particularly noteworthy. This short man sat for exactly 5 hours and during the journey the following sequence of postures can be observed no less than 12 times. Starting well back, legs apart, he gradually slides into slumped posture, sometimes using the arms to prop

himself up. As the arms seem to fail to stop the forward movement of the pelvis, he crosses his knees. Next he stretches his legs forward, and ends up with his body and legs in almost a straight line from the neck down, as near to horizontal as can be. After a very short spell in this position he raises himself back up, only to begin another journey down the slope. Each of these sequences takes between 10 and 20 minutes. [5, 48]

The important point to note about the above observation is that the passenger's behaviour, which was not unlike that of others, was essentially cyclical. External forces acted on the body which required intervention to redress them. The interaction, however, was generally automatic.

The need for seat designers to create seats which help to maintain stability, particularly in a dynamic environment such as a moving train, is clear. At the same time, however, the need for the body to be able to move about to reduce the pressures and muscular fatigue induced by the constraints of seating is also clear. Far from being a question of merely accommodating anthropometric and biomechanical properties of the body, therefore, the design of seats encompasses a variety of other considerations – each of which stem from the person-centred view of ergonomics.

Error corrections

The homoeostatic view of the individual operator's behaviour, which is embodied within a person-centred ergonomic framework, conceives of the individual activity as continually compensating for, and adjusting to, a changing environment. The ergonomics design philosophy, then, is to 'observe' such activities and to use the information to design the environment to meet the need for such activities.

Such a continually varying system, however, could also be conceived in terms of some kind of error correction mechanism, in which an 'error' is defined as any 'movement' away from an optimal path. Departure from optimal thermal environmental conditions, for example, could be conceived as an 'error', which the body attempts to redress using its internal homoeostatic body heat regulating mechanisms. Couched in these terms, the servo-based model of the person has been helpful in the study of operator error and its relationship with skilled behaviour:

> The understanding of human functioning has been greatly advanced in recent years by application of the theory of dynamic control systems. Cybernetics, accepted into general thinking, is usually associated with the feedback' concept. While this is indeed correct, the important point is that it should be *negative feedback*, in the sense of *error-correcting* or at least reducing the effect of errors.

> The concept of 'error'. . .becomes meaningful only if we can specify exactly the *correct* performance of the whole system. Correctness is thus regarded

as the *purpose* of the system and errors are defined as deviations from this purpose. [11, 82]

With this concept in mind, it is important to consider how an 'error' is defined. Unless the operator has some concept of what constitutes 'error' (that is, deviation from a goal or standard) she or he cannot act to reduce it:

> An important logistical point, often overlooked, is how arbitrary it can be to designate something as an error. Before errors can be at all corrected, they must first be accurately detected, which can only be done by comparing output via a loop with the original standard input. Only when the input is known very precisely can error be accurately defined and feedback used to achieve success. The point is that what constitutes success must be known first; only then can error be at all determined. Only when one has acted purposively and set a particular goal, can one speak of success or failure and hence of error. [40, 11]

Although in this model an error-correcting operator is in a sense acting in a closed-loop fashion (with the potential always for the individual variables of purpose, interest, awareness, and so on to open the loop), in many situations the time lag between action and effect – and thus between action and knowledge of success or error – can be large. Branton recognised this in respect to the skills involved in train driving:

> . . .in the railway case, the driver is himself an open-loop, nonlinear, system and also operates open-loop. He controls his train only intermittently, rather than continuously, the whole being in effect a ballistic missile proceeding, so to speak, from one target signal to the next. Whatever feedback he has must be complex and depend on his watch plus his knowledge of the route. His ability to correct errors and make up for lost time is very limited. An error committed now would not be known for some time ahead. Because of this, to me at least, the compelling experience in the driving cab was one of helplessness to influence one's fate. [40, 11]

Such observations, made from the viewpoint of the operator rather than the system, lead to distinct conclusions for ergonomics: that the ergonomically designed environment should present operators with unambiguous information about the effects of their actions in order that the fluctuating error-correcting behaviour can occur most effectively.

Finally, while still discussing the nature of the error-correcting operator, it is important also to consider the effect on the operator of being able to carry out this task effectively. Thus, again taking the person-centred view, individuals have purpose and a sense of responsibility; it is these qualities, and others, which they bring to the situation. Stress can be the result in situations where operators feel unable to correct errors appropriately:

> The idea of 'perceiving the locus of control within the person' necessarily implies the kind of autonomy pursued here, albeit at a level of actual

individual (mental) function, whether or not it is merely imagined. Whether perceived or not, whether believed to be within the person or not, the ultimate control over stress effects on behaviour must be presumed to rest with the person. Who else moves his hand on the joystick or his foot on the brake? Control of an operator's action, not least in the transport context, cannot mean anything other than that a discrete action must be decided and carried out by the individual and no one else. That is the common meaning of individual responsibility in practice. It is applied here explicitly to the initiation of even the smallest muscular movement to control a vehicle. Outsiders, such as engineers, designers and managers, may be able to help or hinder this goal. [34, 500]

Fluctuating awareness

That bodily functions undergo periodic fluctuations in intensity is now well-established. In his person-centred view of ergonomics, Branton sought to understand how such fluctuations affect operators at work, particularly in relation to the build-up of stress in conditions of uncertainty:

> In the endeavour to find some rulefulness in observed behaviour during repetitive and other vigilance-type work, I had often noticed the periodic fluctuations in performance reported by Murrell (1962). I then saw on our filmed records occasions of posture changes and meaningless (non-purposive) behaviour even when no specific work was performed.

> . . .The timing of these incidents led me to consider a possible influence of ultradian, as distinct from circadian, rhythms not only on work performance but also on the general 'stream of consciousness'. What it is exactly that fluctuates rhythmically is not at all clear but, once alerted to their existence, one can find scattered through the literature a great deal of evidence for these variations in physiological functions. [40, 15]

The point to be extracted from this line of thought is not merely that alertness fluctuates, but that these fluctuations are themselves important insofar as people can sense at some level the effects of 'drifting' in and out of different states of awareness. This may lead purposeful and responsible individuals to feel concerned or stressed about their performance:

> It is further suggested that the relative monotony of the situation alone can be inherently stressful to the driver, because it conflicts with the need for alertness in his consciousness of the responsibility to control the train, especially at high speeds. [11, 84]

The stress arises, then, as a result of the operator's 'uncertainty' of what may have happened during the periods of reduced alertness:

> The other reason for sudden spurts of activity is that operators report that they recall having struggled with drowsiness and realised that their 'minds

had wandered' from the task. They knew that there was a time gap during which their mental activity had slipped out of control but are quite uncertain about the duration of that gap. This experience of itself does not produce harmful stress. That does, however, arise as soon as the person becomes aware of the possibility that a mishap may have occurred during the moment of lapsed attention. In other words, it is the awareness of personal responsibility which seems to generate the real stress. If, in addition, the operator is *uncertain*, either about the possible mishap or about his capacity to resolve the urgent problem, anxiety increases. [31, 2]

Returning to the observations about fluctuating postural changes and error corrections, Branton was able to observe these awareness changes and 'mini-panics' using, again, fast film techniques:

In the case of the anaesthetists. . .there was clear evidence of 'micro-sleeps' recurring during prolonged monotony. These 'micro-sleeps' recur with a periodicity of roughly 90–100 minutes (following a basic rest-activity cycle proposed earlier by Kleitman). These were sensed by the anaesthetists via unconscious cues from postural feed-back mechanisms – on return to full consciousness after nodding-off. The results were typical 'mini-panics' as the conscientious dozers were jerked back into action when they realised that the patient might have been deprived of oxygen during the moment of inattention. It is possible to see this phenomenon on speeded-up video recordings. [39, 4]

Branton returns, therefore, to the concepts of purposivity and responsibility. Indeed, he considers that the fluctuations, ameliorated as they are by such individual qualities, may help ergonomists to understand the influence that they have on the operator's behaviour:

Purposive action is distinguished from random behaviour not only by its direction, i.e. to the future, but also by its strength. In this respect, purpose is like those hold-all terms 'motivation' or 'interest'. Indeed, an objective can only be pursued when one has enough interest in it. Now, it has been said that *to sleep is to have lost interest*. Wakefulness is then not only a state of consciousness but also one of more than minimal purposivity, of being interested in whatever one is doing. Whereas boredom is a state in which purposivity is low. Such a state, on the border between wakefulness and sleep, must naturally pervade all psychophysiological parameters and so may make purposivity amenable to research. [40, 15]

Summary

This chapter has illustrated the central focus of the Brantonian, person-centred, view of ergonomics. By identifying the core nature of the human input to the system it is possible to devise often novel ways of measuring it. In particular, the discussion returned to consider the part played by the individual's physiology and the psychological mediating effects of a sense of purpose, responsibility and control. A cybernetic model was also presented of behaviour taking a cyclical, servo-based, form in adaptation to internal and external pressures.

Chapter 5

Summary

Sadly, Paul Branton died before his over-arching view of a person-centred approach to ergonomics could be written with authority. Shortly before his death, with Fernando Leal, he began to prepare a manuscript in which his various thoughts were intended to coalesce, but the depth and breadth of the task which he set himself was such that completion of it by one individual, or even by a pair of individuals, was an ambitious hope. Equally unrealistic would be for others to detail his view in anything more than a superficial manner. The aim of the preceding chapters was to begin to extract the salient features from the Brantonian View of ergonomics and to present them to the reader as a starting point for further thought and development.

The purpose of this final chapter is to summarise the important features of Paul Branton's approach. It will then be left to individual readers to consult his writings for themselves and to extract from them *their own* understanding of the nature of the Brantonian View.

In the previous chapters a view of ergonomics, a particular person-centred view, has been developed which exhibits significant departures from the traditional view of a system. The traditional view sees the system as a closed-loop process in which the two components – 'man' and 'machine' – operate as equal partners. The Brantonian argues that it is the human actor within the system that should play the crucial and dominant role in the safe and efficient operation of that system.

The strength of the Brantonian View, within this person-centred approach, lies in the need to consider not just the person's actions at work, but the purposes which underlie the actions. In doing so, Branton took a 'humanistic' view of behaviour, though using a model which is a much 'harder' form of humanism than the term usually connotes. Thus, he viewed the *person* as central to the ergonomic system, who brings to the situation needs, wishes and desires, as well as the physical and cognitive abilities which are necessary to carry out the task – along with some limitations.

In this conceptualisation, people have a sense of purpose in their work; they approach the task with this purpose 'in mind'. So the system and the task must be designed to accommodate this need. Indeed, one feature of this approach is that the individual's purpose extends beyond just wishing to get the job done. Individuals exhibit activity over and above that required to accomplish the task successfully. Branton demonstrated that this sense of purpose often also involves anticipation and prediction (sometimes even at a quasi-mathematical level) of the consequences, the effects of the action.

It was a crucial matter for him that it is in the nature of workers that they will approach their tasks with a sense of responsibility. This means performing their tasks in the most efficient way and having some kind of control over their actions and that of the system. When these requirements are contravened, when instances occur over which individuals lack control or sense that they have failed, even momentarily, in their responsibility, stress and other negative effects typically occur.

Such are the human needs which are brought to the working system. However, Branton took the person-centred view a few steps further. He emphasised the importance of understanding not only the individual's physiological function and structure but his or her understanding of their situation. He was thus developing a 'theory of mind' – within this person-centred framework.

Considerable weight was placed by him on the importance of the body's various rhythms. Fluctuations in behaviour and attention which take place over time when a person interacts with the system were seen by him to beg explanation. When undertaking relatively boring, though responsible, tasks, for example, workers may develop 'mini-panics' when faced with the 'realisation' that they had 'drifted off' momentarily during the task. Fluctuations in posture helped Branton to understand people's comfort-seeking behaviour, and studies of rapid eye movements were undertaken for insight into such features of work as error susceptibility, monitoring behaviour, decision-making at work, and so on. Changes in EEG patterns, in alpha rhythms in particular, were examined for clues about fluctuations in attention.

While encompassing all of these aspects, the Brantonian View adds an additional factor to the person-centred equation: that of the working person's philosophy and understanding of his or her own situation. Central to this view is the concept of the 'self', including the 'unconscious' dimension. It is this feature which often 'drives' the individual to behave as she or he does; the knowledge needed to perform activities effectively arising, as it were, from 'within' the individual. If we can develop a theory of the mind, then we are in a better position to build a theory of human action.

The Brantonian View is not a solution, however, rather it represents a question; it is not a stricture, rather a structure. The following quotation conveys this 'explanatory model' of behaviour:

> While the foregoing conjectures about natural fluctuations in awareness explain some of the observed behaviour, the self-reports of stress can best be explained by the assumption that operators actually feel responsible for their human charges. They have made it their purpose to conduct themselves safely. A philosophical, meta-psychological case for a purposive explanation of a broader spectrum of human behaviour has been set out

briefly elsewhere [29]. For the present, a subsidiary model is proposed which takes the form of a strictly autonomously controlled person, necessarily possessing value standards and interests in social relations, perpetually seeking and evaluating information from the surroundings. The search varies in intensity, depending on arousal level. This level normally fluctuates in cycles of about 100 minutes. The model concerns the form of mental operations, their contents being material either taken immediately from the surrounding world or from stored, primarily emotive experiences. The explanation is purposive, rather than causal, as it is argued that the thoughts which determine behaviour are forecasts of future states of affairs and their consequences for the person, rather than past experiences in themselves of speculative origin. Needless to say, this statement can only be the briefest of telegram-style sketches in the space available. [34, 505]

It is now up to others to take this 'telegram-style sketch' of the person who is at the centre of the system and develop it into a more comprehensive philosophy for ergonomics theory and practice.

References

Bruner, J. S., 1964, The course of cognitive growth, *American Psychologist*, **19**, 1.

Bruner, J. S., 1980, *Beyond the Information Given*, London: Allen & Unwin.

Davis, D. R., 1966, Railway signals passed at danger: The drivers, circumstances and psychological processes, *Ergonomics*, **9**, 211.

Eason, K. D. 1991, Ergonomic perspectives on advances in human-computer interaction, *Ergonomics*, **34**, 721–742.

Murrell, K. F. H., 1962, Operator variability and its industrial consequence, *International Journal of Production Research*, **1**, 39.

Murrell, K. F. H., 1980, Occupational psychology through autobiography: Hywel Murrell, *Journal of Occupational Psychology*, **53**, 281–290.

Polyani, M., 1958, *Personal Knowledge*, Chicago: University Press.

Shipley, P. and Leal, F., 1991, *The active self: Beyond Dualism*, paper presented to Annual Conference of the British Psychological Society, in *Newsletter of the History and Philosophy of Psychology Section of the BPS*, November 1991.

Singleton, W. T., 1967, Acquisition of Skill – The theory behind training design, in Robinson J. and Barnes, N. (Eds), *New Media and Methods in Industrial Training*, London: BBC.

Taylor, C., 1964, *The Explanation of Behaviour*, London: Routledge & Kegan Paul.

Teichener, W. H., 1967, The subjective response to the thermal environment, *Human Factors*, **9**, 497.

Wisner, A., 1989, Fatigue and human reliability revisited in the light of ergonomics and work psychopathology, *Ergonomics*, **32**, 891–898.

Wright, P., 1986, Phenomena, function and design: Does information make a difference?, *Contemporary Ergonomics 1986*, Oborne, D. J., (Ed), London: Taylor & Francis.

Part II
The Brantonian Contribution

Chapter 6

Paul Branton as a philosopher

Fernando Leal

University of Guadalajara, Mexico

Paul Branton understood himself as an heir of the psychologically oriented tradition in critical philosophy.[1] According to this tradition, all human knowledge and action is 'grounded' or 'based' upon some original and basic mental capacities. To philosophize is to find out what these capacities are and to prove their existence. A critical philosopher is actually an applied psychologist.[2] The traditional, although nowadays increasingly unfashionable way to express this, is by talking about science and its metaphysical foundations.[3] The term 'metaphysical' is of course immaterial; and I only keep it because Paul was fond of it. Now if we use an F to signify the foundations underlying science (S), a flow chart can represent the essence of the critical vision as in Figure 6.1.

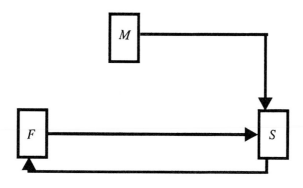

Figure 6.1

We can then say that the arrow which leads from F forward to S represents a *foundational* relation, whereas the arrow which leads from S back to F represents a *critical* one, critique being the (scientific) activity of finding out what is in the black box F which makes S possible. By the way, Figure 6.1 presents an important complication, which lack of space prevents me from saying much about: mathematics (M) is assigned by critical philosophy a very special – not foundational, but nonetheless indispensable – role in this scheme.[4] Even today there is no consensus as to what is the relation between mathematics and the scientific disciplines, e.g. physics. Some argue that its role is only *instrumental* (it helps give scientific findings a perspicuous shape and allows for fast calculations), whereas others think that it is *constitutive* of science.[5] It is impossible to do justice to that debate here; the best policy is therefore to avoid giving any name to the relation represented by the arrow leading from M down to S.

The above blocks are black boxes, so what is inside them? Within the tradition of psychological critique, there is only one sustained attempt at opening up those black boxes, that of J. F. Fries.[6] His starting point, like that of all nineteenth-century scholars is to distinguish between the study of nature and the study of mind. For the purposes of a diagram let's assign a number to those two groups: S1 should then be the natural sciences and S2 the sciences of the mind. We are not interested here in the contents of S1. As to S2 Fries proposed a very interesting subdivision, his arguments being that when we study mind we *either* study it in relation to its own internal states (general psychology) *or* in relation to material things (the 'pragmatic' or engineering sciences) or else we study the interaction of minds (the social sciences proper).[7] Again with our diagram in view, let's call these the *a*, *b* and *c* levels of the study of mind. For example, the label 'S2*c*' represents the social sciences. Now, according to Fries, although all sciences share some common metaphysical foundations, each group and each level has some particular foundations of its own. Thus, to the box S2*c* will correspond a foundational box F2*c*, and so on.

The last paragraph is pure Fries. But Paul went a step further on. Being an ergonomist, he was particularly interested in the place his field should have in the general scheme. I believe his work suggests that he conceived of ergonomics as a kind of mediator at the *b* level, i.e. a mediator between metaphysics and engineering. In a moment we shall see how. But before this, I want to say that Paul seems to have asked himself another question: if there is some mediation at the *b* level, what happens at the *c* level? In other words, what discipline might be able to mediate between metaphysics and the social sciences? In an unpublished manuscript, Paul called this putative discipline the 'social psychology of the individual' and intimated that it does not exist yet as such, but that it was badly needed. When we try to represent all this in a flow chart, something like Figure 6.2 emerges. This expansion of the original block diagram seems to me a faithful if simplified representation of the structure which was in Paul's mind when he tried to think about critical philosophy. Please notice that feedback from the scientific box to the metaphysical box does not only start from general psychology (S2*a*), but also from the mediating psychological disciplines ('E' being shorthand for ergonomics and 'X' for the still undeveloped mediator between metaphysics and the social sciences).

When trying to outline the contents of the foundations of the study of mind at the *a*, *b* and *c* levels, Paul talked about 'the three theoretical frameworks of the theory of reason' as follows:

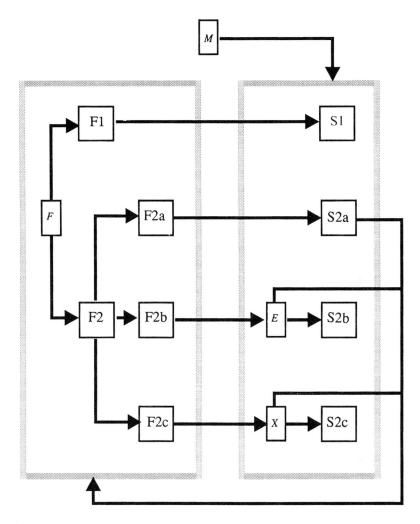

Figure 6.2

The Self must be represented to itself [F2a], and the Self must relate itself to the outside world and test its 'reality'. This relating takes two forms: *as a ship at sea the Self must orientate itself in the spatial world* [F2b], and *it must locate, relate and engage itself as a bearer of interests, the Ego, in the social world* [F2c].[8]

One can think of these statements as the metapsychological principles of a well-defined axiomatic psychology, or one can see them as the basic capacities of the human mind, i.e. those capacities that make possible for people to know and do things.[9] However, let us not forget that Figure 6.2 is only an ideal. All human knowledge and action is somehow contained in that figure, which is quite a mouthful, even for a philosopher. But, although the diagram is obviously very schematic, it might be helpful, especially if we use it as a map of the Brantonian territory. In this map we can distinguish at least ten different 'thinking tracks'; to see them clearly, we can erase the arrows marking the flow from one block to the next. The map of Paul's wanderings looks then like Figure 6.3. As the reader can appreciate, he was quite a walker – in a manner befitting someone with a Germanic background. The numbers correspond to what I take was the chronological order of his intellectual development, although there must be of course some overlaps.[10] The following sketches are intended as preliminary descriptions of each track.

1. **From S2a to F2a, or from general psychology to its metaphysical principles.** If I should have to summarize this part of the Brantonian approach in a single statement, the best would probably be this: *the human mind is internally self-active.* Already as an undergraduate student of psychology, especially the psychology of perception, Paul had brooded over a puzzle: How is it possible that our visual perception is continuous, given that we are moving all the time? When he said 'moving', he not only meant the relatively slow movements of our limbs or eyes, but also, and especially, the very rapid micro-movements of the eyes, the so-called REMs which take place even when we 'fixate' the eyes and are otherwise still. We do not see a succession of discrete pictures, but a continuous world, although there are actual interruptions occurring all the time. Therefore, an enormous amount of constructive activity must take place inside us. Or rather, according to Paul, this activity does not 'take place inside us', but *we do it ourselves.* Paul also knew, as one who actively participated in Ditchburn-like experiments, that if you stop the eyes from moving, you stop the seeing altogether. He boldly concluded (Branton, 1984) that those actual physical interruptions produced by our own movements should not be seen as obstacles to continuous perception, i.e. to perception as we know it, but as necessary, or to express it in critical jargon, a 'condition of possibility' of visual perception. But if so, they are then a metaphysical foundation, part of the system of principles of general psychology (S2a).

2. **From E to F2b, or from ergonomics to its metaphysical principles.** Afterwards, and in connection with his work on ergonomics, Paul would learn that there are other micromovements taking place in other sense organs (like tinnitus in the inner ear) and as a matter of fact in the whole muscle system ('muscle tremor'). He also appreciated that there are many so-called *oscillators* and *rhythms* at all levels. And so he began to think of a generalization; that one could build a general theory of all self-initiated physiological changes as underlying ('making possible') perception, dreaming and action. Nobody would listen to him; when he once approached a famous professor with the question whether REMs might be related to tinnitus or even to muscle tremor, the reply was 'I am not interested; all that is outside my field'. Since Paul had not the means to conduct the appropriate experiments, many of his ideas remained speculative.[11] One of them, however, deserves special mention: in his ergonomic work he was always insisting that people never know with certainty what is going to happen next. As he quite metaphysically put it, the world is uncertain.[12] But the important thing is not to rest content with *that* assertion, as would be the case in traditional metaphysics, but to make the decisive critical (psychological) step, viz. to say that, although the world is in itself uncertain, *the uncertainty of the world is reduced by human activity.* This is, of course, a common idea in human factors; but its enormous metaphysical import is not really appreciated, partly because of the division of labour between philosophers and scientists, and partly because uncertainty reduction is usually only taken seriously at macrolevels. As against this, Paul

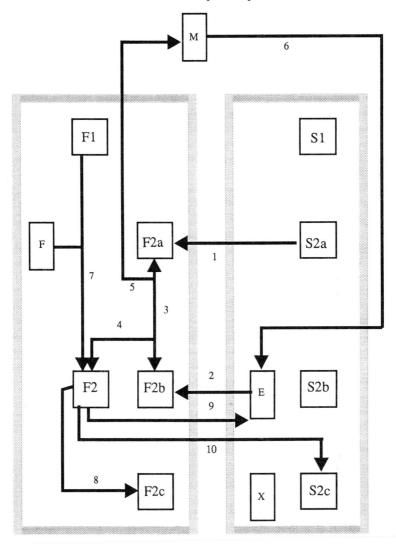

Figure 6.3

thought that uncertainty reduction goes all the way down to micromovements and oscillators. For him, it is we ourselves who make all these inside movements and thus reduce uncertainty. As Paul was fond of putting it, 'we make our own waves and float on them'; in particular, it is by means of those waves that we manage to distinguish self and non-self. *For only what I do* (e.g. REMs) *is certain, i.e. certainly produced by myself.* Everything else is uncertain. Uncertainty reduction through self-activity occurs at all levels and it is the underlying principle of perception and action.[13] In fact, the core of free will might be located right here.

3. **From F2b to F2a and back, or from the metaphysics of ergonomics to the metaphysics of psychology.** The metaphysical principle of self-activity and the metaphysical principle of uncertainty reduction are clearly connected. This connection

had launched Paul into what we might call the metaphysics of human action and even 'the metaphysics of work'. He started looking for other basic features of human action. His analysis of skills and tasks, particularly of the train driver, led him to the discovery of one such, viz. *that human action is oriented towards the future.* This discovery has enormous consequences. It is very important to emphasize here that Paul was not just thinking of 'making plans for the future' or of 'having intentions'. These things exist. And they are certainly very interesting, especially for philosophers sitting in their armchairs. But Paul meant business. Take train drivers (Branton, 1978, 1979). They have to drive the train; no time here to make plans and have intentions. Train drivers relate to the immediate future. As Paul was fond of saying, the train is actually a gigantic missile, which is being shot at enormous speed and weighs several thousands of tons. And insisting on the place of moral values in every work situation, he added that the health, safety, comfort and welfare, nay the very lives of a great number of people depend on the skill of the drivers. For example, stopping the train takes several miles and a considerable amount of time. The decision to apply the brakes must be made at a point in time when the train driver cannot see where the train is going to stop. (This is why trains are so good an example; similar things happen also when driving a car, but the spatial and temporal span is so much smaller.) This 'acting ahead of time' is thus a farther metaphysical principle or basic human capacity.[14]

4. **From F2*ab* to F2, or from the combined metaphysics of psychology and ergonomics to even more general metaphysical principles**. It was more or less at this point that Paul read an important essay within the tradition of critical philosophy, Professor Grete Henry-Hermann's essay 'Conquering Chance'.[15] In that essay, she concludes *inter alia* that his teacher, Leonard Nelson, made a mistake in thinking that when someone acts deliberately, i.e. for a reason, the deliberate act is caused by the reason which prompted it. The whole analysis of human action which we find in Nelson is causalistic. Professor Henry-Hermann's own research on the philosophical foundations of quantum mechanics (to be located in box F1 of Figure 6.2) had forced her to doubt the universal validity of the causal law. In particular, she pleaded against its application to our inner life, proposing a distinction between 'being causally determined' and 'being determined by reasons'. It is interesting that analytic philosophers in Britain were making similar proposals at about the same time.[16]

One of the things which is different in Henry-Hermann's approach (as opposed to analytic philosophy) is that she is asking how an action can be *determined* whereas analytic philosophers prefer to ask how an action can be *explained*. This difference can be bridged without difficulty. To avoid misunderstanding in the Anglo-Saxon world, Paul used to talk about the explanation of human action, although he occasionally lapsed into the older determination-talk. But there is a more interesting problem: both Henry-Hermann and analytic philosophers talk about *reasons* as being that which determine or explain human action. Paul found reason-talk both too intellectual and too weak. He preferred to talk more directly about *purposes*. The word 'purpose' is not perfect, either, and it has several undesirable connotations, but it is better than the word 'reason'. It is also more common in psychological contexts. One of the troubles with the word 'purpose' (which it shares, by the way, with 'reason') is of course that it has come under fire from many sides, because it is allegedly teleological; and teleological explanations are rejected by practically every scientist. I do not want to go into that complex debate, but I think that Paul's ideas about the 'orientation towards the future' of all human action can contribute to the construction of a scientific concept of purpose. Nelson's concept of 'conquering chance' (which had been so heavily criticized by Grete Hermann) would have to be replaced by Paul's concept of 'conquering the future' – which is what internal representations are all about. Such a conquest can only take place if people have purposes; for purposes are representations of what should be (in a not-yet-ethical sense of that expression). And although Paul was not able to work out the form of a purposive explanation to his own satisfaction (see Branton, 1981, 1983, 1985, 1989), he certainly thought that the principle of such explanations belong to what Fries would have called the 'metaphysics of internal nature' (block F2 in Figure 6.2 and 6.3).

5. **From F2*ab* to M, or from the combined metaphysics of ergonomics and psychology to mathematics.** Take train drivers again. What they achieve is astonishingly precise both in space and in time. In fact, it is the equivalent of complex operations in infinitesimal calculus. There is of course not enough time to differentiate and integrate on paper. And conscious mental operations have to be ruled out, too, given that train drivers haven't studied any higher maths. What on earth goes on in their heads then? Whatever it is, it isn't conscious, although it is done with total certainty and accuracy. Paul was fond of talking here of *quasi-mathematical operations and representations*. Over the years, he kept collecting papers written by other people which substantiate the need for this kind of representation, although these other people were either as puzzled as Paul or even unaware of the problem. What are quasi-mathematical representations? Paul was never clear about them. He thought that they are certainly not symbolic and not algorithmic, which confirmed his old suspicion that the 'computer metaphor', so cherished by cognitive scientists, was totally misleading.[17] If these inner processes are not symbolic, what are they? Paul tried to apply a concept which Jerome Bruner had taken from William James – the concept of *enactive* representations, i.e. representations which are inherent in human action, embodied in a skill. According to Bruner, such representations are different from both iconic and symbolic ones, i.e. they are neither like images nor like words. That's what makes it so difficult to picture them or to talk about them. But they must exist, and they are immensely relevant for action. And they by right belong to a complete system of metaphysical principles as well. It may be worth mentioning here that there are some recent attempts to understand the processes, both within psychology and within the philosophy of mathematics, according to which all of mathematics originates in pre-scientific human action, work and technology.[18]

6. **From M to E, or from mathematics to ergonomics.** This would be the place to say something about the application of mathematical methods to psychology and ergonomics, and especially about measurement and statistics. Paul's colleagues admired his ingenuity in devising novel ways to define elusive qualities (say, discomfort), so as to be able to measure them. And he certainly thought that we must try to measure wherever this is feasible. I must refer the reader here to Dave Oborne's contribution to this volume.

7. **From F1 to F2, or from the metaphysics of nature to the metaphysics of people (and thus to general metaphysics).** Paul's interest in purposes led him to quantum mechanics. He read widely about this theory and had many discussions with physicists. Although the whole field is very complex, I would like to mention two things which gave Paul ample food for thought. Firstly, quantum theory implies that the causal law (as mentioned above) is restricted in its application. This is important because strong determinism depends on the universal validity of the causal law; and determinism has often been used to prove that the moral law is not valid, because we wouldn't be free. Under the dispensation of quantum theory, the reality of free will (and thus of moral values) is not proved, but it is made less unreasonable. According to Paul, we should distinguish three domains: the *undetermined* events of quantum physics, the *causally determined* events of deterministic (classical and relativist) physics and the *self-determined* events of purposive action.[19]

All this is hardly original, or only so when the details about purposivity and 'conquest of the future' are appropriately spelt out. But the second idea is intriguing in itself. Paul was inspired to postulate the existence of what he called the 'psychological quantum'. According to him, something like this had to be admitted as a principle in order to understand the act of knowing (think of the *Aha-Erlebnis* or sudden flash of insight), and particularly in order to understand the act of abstraction within ethical reasoning which precedes deliberate action in a situation where two interests are in conflict.

8. **From F1 to F2c, or from the general metaphysics of people to 'moral' metaphysics.** The idea of the psychological quantum brings us to Paul's real starting point, *viz.* ethics. At the moral and social level of the system of metaphysical principles for psychology, he thought that one had to postulate something inside us which accounted for the 'good emotions'. He shared with the critical tradition the idea that intellectual conviction ('I ought to act in this way now') was not enough, at least if unaccompanied by an appropriate feeling or emotion which would do the motivating work. Thus was born the idea of the so-called *moral interest*.[20] Paul was interested in a possible physiological basis of such a moral interest. And he thought that our growing knowledge about hormones and neurotransmitters would be relevant here; he was looking for what he called the 'sociotropic' components of our emotional life. Although this is again a field which bristles with conceptual and empirical difficulties, I must mention it because it was immensely important for Paul.

    On the other hand, the main feature of interests in the critical tradition, what distinguishes them from all other mental processes, is that they are *valenced* (this is clearest in emotions like love and hate). And so Paul was always looking for evidence of oppositions which might support such valencing. In particular, he thought that the opposition of agonistic (or synergic) *vs.* antagonistic muscles, and maybe even the more general opposition of stimulation and inhibition in the nervous system, might be such a physiological basis.

9. **From F2 to E, or from the metaphysics of people to ergonomics.** We have been getting progressively nearer to traditional ethics. Contemporary analytical philosophy distinguishes two fields of research: ethics or moral philosophy on the one hand, and the philosophical *theory of action* on the other. According to this school the latter precedes the former and is actually independent of it. But traditionally both things are conflated. In a sense then, we have been moving from the theory of action towards ethics (although of course I did mention the responsibility of the train driver in a previous paragraph). Paul saw early in his career that general psychology promotes a false conception of human beings. The defects of this conception are often unclear when we are theorizing in the abstract, but have serious consequences when they are applied in practice – in medicine, psychotherapy and ergonomics. Paul fought again and again for workers to be recognized and respected as autonomous, feeling, active, purposive, valuing and valuable beings. Thus, he pleaded in international committees for giving priority to health, safety and welfare over efficiency. Such debates are familiar in ergonomics and human factors, and also a good example of *conflict between human beings*. Paul was particularly interested in the philosophical principles underlying the *resolution* of conflict. He believed that Nelson's analysis of the kind of reasoning – ethical reasoning – which is needed here was essentially correct and he very much admired Nelson's version of the moral law, *viz.* 'Never act in such a way that you could not also assent to your action if the interests of those affected by it were also your own'; or more simply: 'Act as though the interests of those affected by your actions were also your own.'[21]

    This crisp formulation was a positive improvement over the cumbrous wordings of both Kant and Fries, but totally faithful to their intentions. Paul asked himself as a psychologist: if such is the moral law, then what are the necessary psychological operations people should be able to execute in order to apply it in the real world. This question belongs to the theory of practical reason. It asks about what people should have to be like – what they should be capable of – if the moral law has any reality at all. If people are not like that, if they are not capable of carrying out the necessary steps of ethical reasoning, then the moral law is only a dream, a figment of our imaginations, a pious thought. To say it in Kantian idiom, Paul wanted to know the psychological conditions of possibility of the moral law. After an extended analysis Paul thought that a five-stage reasoning process had to be postulated:

1. *Sympathy*: 'recognise that interests of persons other than yourself are involved and find out what they are'.
2. *Empathy*: 'put yourself into the other person's shoes; introject their interests into yourself and compare them with yours'.
3. *Abstraction*: 'detach interests from whoever holds them and assign intersubjective value to each'.
4. *Weighing-up*: 'imagine possible outcomes if either one of the conflicting interests prevailed and were to become real and general in future'.
5. *Wilful decision and action*: 'actively prefer one outcome and make it become real'.

According to Paul, there is a gap between socially skilled acts (corresponding to stages 1 and 2) and ethical acts proper (stages 3 through 5). To get over the gap one has to jump. And Paul thought that it is useful to think of that jump as similar to the quantum jump of physical theory.

In a sense, Paul's analysis is not strikingly original. But then it was not intended as original. When people started criticizing Kant by saying that his categorical imperative was old hat, he replied that it of course was, that he was not such a fool as to pretend to teach humankind, from scratch as it were, how they ought to behave.[22] What Kant wanted was just to make a little clearer, to articulate, what people already know. Something similar applies to Paul's analysis. But maybe a couple of points should be emphasized to ascertain its utility. On the one hand, philosophical accounts of ethical reasoning (such as Nelson's) remain hypothetical as long as there is no empirically grounded psychological theory which effectively demonstrates the reality of the operations and underlying capacities implied in that reasoning. On the other hand, psychological studies usually select one or two aspects of ethical reasoning and fail to see the relevant facts in relation to each other. What is missing in the philosophy might thus be supplied by psychology and vice versa. A possible piece of research might illuminate what I mean. Psychiatry, psychotherapy and psychopathology have to do with people who manifest abnormal reasoning and deviant behaviour, e.g. sadists and psychopaths. Since Paul's analysis of ethical reasoning contains an ordered ranking of psychological functions, it could be used to construct a typology of the corresponding psychological dysfunctions. If normal human beings are able to perform the operations indicated in Paul's analysis, a classification of incapacities which underlie different mental pathologies becomes possible. For instance, the usual description of psychopath corresponds roughly to Paul's first stage (it is lack of sympathy), whereas the sadist would be normal at stage 1 but abnormal at stage 2 (sadism is rather lack of empathy in Paul's sense). I think this is a fruitful idea which should be followed up, and as a matter of fact the study of the psychopathological literature from that point of view might help improve Paul's analysis as well.[23]

10. **From F2 to S2c, or from the metaphysics of people to politics.** The last of these brief sketches is an end which is also a beginning. Paul had always a keen interest in politics, long before he thought of studying psychology. He was an undogmatic socialist; in fact he always abhorred Marxism, not least because of the deterministic element in it. Probably his best essay on the subject is 'On being reasonable', written in honour of a German friend (Branton, 1981). He liked that English expression, *reasonable*, and thought it was a much better ideal than cold rationality. Although he referred to the book he tried to write after retirement under several names, the one he adopted in the end was *A Psychology of Reasonable Autonomy*. The

autonomy he wanted to establish for people was guided by reason, but not by instrumental reason, blindly bound to machine-like efficiency. For Paul, reasonableness belongs rather to the fundamental idea of *citizenship*. His commitment to the worker belongs to this strand of his thought. All along his career as an ergonomist, he tried hard to find ways of improving the conditions under which people work and of avoiding so much unnecessary human misery, illness and so-called 'accidents', both for the workers themselves and for the consumers. And he thought that the kind of careful and committed thinking a good ergonomist brings to bear on these issues has to be extended to the whole of society. A slogan for that might be: *People should always be treated as reasonable beings.*

We are now at the end of what is only the barest of skeletons of Paul Branton's philosophical thinking. As I suggested at the beginning, the need for an orderly exposition has forced me to present Paul's ideas in the framework of a philosophy of science. At the end of this exposition, I must hasten to add that, to use Kant's distinction, Paul was more interested in *practical* than in *theoretical* reason. He was thus not so much trying to lay the foundations of (psychological) science as attempting to establish the conditions of possibility of ordered, skilled, socially integrated, and responsible human action. Continuous self-originated activity and physiological rhythms, reduction of uncertainty, the conquest of the future, purposive explanations, the construction and criticism of values, resolution of conflict by reasoning and reasonableness, normal responsible behaviour of human operators, quasi-mathematical operations – although these and other Brantonian concepts still require development, we must be grateful to Paul Branton for his untiring insistence on their importance and fruitfulness.[24]

Notes

1 The most important names in this tradition are Jakob Friedrich Fries, Ernst Friedrich Apelt and Leonard Nelson, who claimed to be followers of David Hume and Immanuel Kant. A partial historical account of that tradition is Leonard Nelson, 1970 (Vol. 1), 1971 (Vol. 2), *Progress and Regress in Philosophy*, Oxford, Basil Blackwell.
2 Of course, in the course of the philosophical search we have to use other techniques, e.g. logical techniques, but this is hardly exclusive to philosophy.
3 Paul's thinking shows an unresolved tension here, since he was rather more interested in action (practice) than in knowledge (theory, science) and his work contains the germ of a conception according to which action is prior to knowledge. Pat Shipley and I have begun to spell out that conception (see our 'The Active Self', *Newsletter of the History and Philosophy of Psychology Section of the British Psychological Society*, November 1991). In the meantime, it is easier to present Paul's ideas in the traditional framework of a philosophy of science.
4 Already Kant had stated that a discipline can be reckoned to be scientific only as far as it has been mathematized (*Metaphysical Principles of Physics*, 1786, Preface).
5 For the latter view see Hilary Putnam, 1975, *Mathematics, Matter and Method* (Cambridge University Press); for the former, Hartry Field, 1980, *Science without Numbers*: Princeton University Press.
6 See J. F. Fries, 1824, *System der Metaphysik*, Heidelberg: C. F. Winter.
7 It is noteworthy that for Fries, the engineering sciences are sciences of the mind. I cannot discuss his arguments here, but maybe the rest of the chapter will help show that it is not a wildly implausible idea. Fries's ideas about the organization of the sciences are, needless to say, vastly more complicated than I am able to represent here.
8 This quotation is taken from the unpublished manuscript already referred to.
9 Notice the modal verb 'must' in the three statements above: in Paul's idiom that verb

always indicates a metaphysical necessity, a 'condition of possibility', a postulated basic human capacity.

10 Thus, an early manuscript (Branton, 1960) suggests that the ideas corresponding roughly to tracks 8 and 9 were already present in the beginning. As a matter of fact, politics (and therefore track 10) was what originally made Paul tick: it seems that it was the need for a reasonable political order which prompted him to study psychology in the first place. In a sense, Paul's work can be best understood as trying to close a loop started in his youth.

11 But not all. Some related ideas were exploited in his ergonomic work. This work and generally the importance of physiological rhythms for the Brantonian approach is highlighted in Dave Oborne's contribution to this volume.

12 This statement is less metaphysical (in the traditional sense) than it seems and so a little misleading. Being a psychologist, Paul was rather more interested in the *psychological* uncertainty than in any *inherent* or *ontological* uncertainty of the world, as has become popular from quantum physics (but not only; see Patrick Suppes, 1984, *Probabilistic Metaphysics*, Oxford: Blackwell).

13 This enormously innovative idea is very difficult to appreciate because people usually equate certainty and knowledge with consciousness and awareness. But this is a mistake. See 'The Active Self', ref. 3 above.

14 Paul's discovery goes against the orthodox idea of the human mind 'living in the present'. According to Paul, it rather lives in the future, especially in the immediate future.

15 Originally published in German in 1953. At Paul's initiative, this essay has been now translated into English and published in a British philosophical journal (*Philosophical Investigations*, vol. 14, no. 1, January 1991, pp. 1–80).

16 Most prominently among them Gilbert Ryle in Oxford (*The Concept of Mind*, 1949, London: Hutchinson) and Ludwig Wittgenstein in Cambridge (1945, *Philosophical Investigations*, Oxford: Blackwell). There are several reasons why these philosophers, and especially Wittgenstein, had lots of followers, whereas Professor Henry-Hermann didn't; I regret I cannot go into them here. In more recent philosophy, the 'causal theory of action' has been powerfully resurrected by Donald Davidson and his disciples; but again lack of space prevents me from discussing any of this.

17 A famous mathematician has recently published a book, one of whose central arguments is that mathematics in people's heads (including mathematicians' heads) are something totally different from mathematics on paper. See Roger Penrose, 1989, *The Emperor's New Mind: Concerning Computers, Minds, and the Laws of Physics*, Oxford: University Press.

18 Jerome Gibson's theory of perception through movement is well-known, although few people mention his admittedly sketchy ideas about the origins of geometrical space. Philip Kitcher is a philosopher in the USA who insists that all mathematics is grounded in ordinary human practices. And the so-called Erlangen school in Germany (headed by Paul Lorenzen) has developed a whole research programme to investigate how abstract theories in logic and mathematics are related to and indeed dependent on the creation of tools and the manipulation of the environment.

19 According to recent work by H. H. Rosenbrock (1990, *Machines with a purpose*, Oxford University Press), the causal and purposive modes of explanation are mathematically equivalent. His analysis applies to both classical and quantum physics. This looks like a complete breakthrough, but I am still not in a position to pass any judgement. But if it is true, a modified version of Paul's ideas might be developed.

20 The word 'interest' is a semi-technical term of critical philosophy for the internal act of valuing. As such it is opposed to knowledge and action; in fact it is a kind of mediator between both. The best exposition within critical philosophy of the general conception of interest is in Leonard Nelson, *Kritik der praktischen Vernunft*

(Göttingen, 1917), Part III, Section 1; for the moral interest, see Section 4. The word was later used in the American version of Austrian 'value theory' (cf. R. B. Parry, 1926, *General Theory of Value*, Cambridge, Mass.).

21 Leonard Nelson, 1917, *Kritik der praktischen Vernunft*, p. 133, Leipzig: Veit & Co.
22 Cf. Kant, *Critique of Practical Reason*, Preface, Footnote 3.
23 This wouldn't be dissimilar to the philosophical illumination obtained from neurology and neurosurgery, as set forth in the writings of Oliver Sacks.
24 The research on which this chapter is based was undertaken during the academic year 1990–91 under the auspices of three institutions: the Society for the Furtherance of the Critical Philosophy (London), which conceded me a generous grant and lots of moral support; by the University of Guadalajara (Mexico), which gave me a leave of absence; and by the Department of Philosophy of Birkbeck College, which provided me with an academic home, even though they didn't quite understand what I was doing. To all of them I want to express my sincerest thanks.

Chapter 7

Thinking is very far from knowing

Nigel Corlett

Professor of Production Engineering, Institute for Occupational Ergonomics, University of Nottingham

This eighteenth century proverb culled, with no erudition on my part from a dictionary of quotations, expresses a truth important for ergonomists, and emphasises a view that Paul Branton espoused. The view is that there is a realm of 'unconscious' knowledge on which any expert, skilled person relies, but which is not always consciously available.

Young ergonomists fresh from college may enter their first appointment with their minds busy with methods, approaches and knowledge. Their training has told them that a problem has to have an appropriate scientific approach, formal analysis and formal conclusions. The nuances of the problem may well pass them by and the 'smell' of the situation not noticed. They read the words, but do not hear the music.

All of this is a preliminary to arguing that ergonomics, as a subject, is at risk of being defined as less than the sum of its parts. The love of theory and the perception – at the very least British – that *thinking* is superior to *doing*, that thinkers are superior to doers, puts a premium on research over practice. Even individuals who have demonstrated abilities in both areas have to plead publicly for forcing the attention of researchers onto the utility of what they do. Chapanis (1991) argued that all published papers in the subject should be required to carry design recommendations arising from the findings. In this way, he argued, results would be more readily introduced into practice, and researchers' minds better focused on the reality of what they do.

Clinging onto a belief that one form of exercising the human organism has superiority over another form – rather than being just different – is a subject for discussion in its own right, too extensive for presentation here. But it is a belief which should be considered carefully since, unconsciously or otherwise, it biases our attitudes to what we do and how we do it.

If one reviews the papers submitted to a professional journal it is striking how few go beyond the investigation of a segment of a problem on a limited number of dimensions. There is a need, of course, for studies which do the equivalent of assessing how far apart controls should be to permit a user to reach them with a certain probability of success. These are the building bricks of the subject. But the step of testing results beyond the laboratory is rarely taken. Why this is so is not difficult to surmise. The programme is difficult to do, it is often difficult to find suitable test sites, and even where it has been possible, the research sponsor industry finds the interference arising from field testing is another problem which it can easily avoid.

Yet the lack of published work where the contribution of ergonomics in a working environment is considered from a broad perspective, is notable. There is no allocation of blame here, we are all guilty if guilt must be apportioned, though it is really counter-productive to do so. What we must seek is ways to extend the view of what constitutes an ergonomics investigation so that more studies embrace the measurement of the major consequences of implementation. Whilst attempting such a positive action, some negative action would also be appropriate. It would be desirable to cut down to size the too-widely held view of academic research as a process of literature survey, experiment, statistical analysis and conclusions from that analysis. These may well be aspects of a research study, but so often ignored are the limitations which arise from the choice of a few variables, the validity of the results in relation to the experiment's size and design, the role of statistics in defining not the reality of conclusions but the probability of differences between conditions; such limits to experiments are not often discussed in papers. How often do the authors step back far enough from their work to view the wider context?

There is a great need to merge theory and practice more intimately so that they inform each other, to encourage the intermingling of both the art and the science of ergonomics, and to encourage more interaction between the investigator and the investigated. To depend upon the science alone is to assume we know more than we do, whilst ignoring the vast areas of understanding surrounding real world situations. This is no new idea, of course. Studies of process plant controllers revealed that they did not spend all their time looking at the control panels, but used a far wider range of information. Even the simple act of sitting showed that the requirements of the seated person were not fulfilled by applying theories of seat design; see for example the investigation by Shackel, Chidsey and Shipley (1969), and Paul Branton's own analysis of railway passenger sitting behaviour.

But the belief that thinking is superior to doing, and therefore that thinkers are superior to doers, leads to more serious consequences. The separation between those who get their hands dirty and those who do not runs back through centuries, and is one of the foundations for the caste system of many countries, for example, and the down-grading of the artisan in many industrial relations structures. It is inappropriate to discuss politicians extensively here, but some comment is relevant, with reference to the British scene. There is only a handful of members of the British parliament who are engineers, craftsmen or similar. Almost all are bankers, lawyers, accountants or people with a 'good' education but little experience in scientific or practical, technical matters. Their experience is often used in argument in support of their political beliefs,[1] and the implication is that argument, or increasingly in recent years, assertion, creates the facts. Too often, evidence in support is selectively used; too often the subsequent decisions affect a wider field than just the political. Competent civil servants may be available to advise, but ministers can, and do, ignore such advice if it is politically inconvenient.

Although history presents the lessons, they are not always learnt. The North aisle of St Paul's Cathedral, in London, carries a memorial to the entire complement of 'HMS *Captain*' which sank in the late nineteenth century. The conclusions of the official enquiry are engraved on the brass plate. They are quite unequivocal. The House of Commons insisted that the load line was raised, in spite of protests from the designers. The result was catastrophic. Contempt for technical expertise against the judgement of 'great men' is implied from such a tragedy. Yet not ten years ago it happened again, when the non-technically qualified management of a ferry company imposed an unsafe work routine against the advice of their mariners. The consequence was the capsize of a ferry, with grave loss of life.

These examples of the wider consequences of a belief in the greater validity of decisions arrived at by theory rather than by informed practice, a view which infects the legal system and much of current management, is one which the ergonomist must avoid. Theory is the consequence of practice, just as knowledge is of experience, and the central focus of ergonomics is the individual at work. Since the consequences of good ergonomics are likely to increase the power of people in relation to their work, ergonomics is also a political activity. Power is redistributed within the work environment. So we see another facet of the ergonomists' activity, and recognise that changing the distribution of power requires the ergonomist to have a perspective on where he or she stands. What are the ethics of and politics of ergonomists?

Because of this subject's multi-factorial nature the ergonomist is in principle open to information from all sources. Such information must arrive in the terms of, and from the perspective of the provider of that information. To impose the investigator's framework may be inappropriate, as may be the assumption held by the investigator that she/he is a servant of management because management is paying.

There is no doubt that sometimes management will find the reports of specialists unpalatable, and will in some cases try to argue for different conclusions more convenient to its aims. It can, for example, get an expert to report that a primary move should be the introduction of training in lifting and handling, whilst ergonomic changes are played down. This could be seen by it as the cheaper option, but it would put the onus for any injury onto the handler. Ergonomists would recognise the decision as unergonomic and the ethically-minded professional would be alert to the dangers of such a situation.

The ergonomist's position is that of an integrator, with a range of specialist knowledge available to assist in the achievement of a better working situation. To get people involved in the solutions to their own problems is more than a move to ease the introduction of new ideas. It is to recognise the contributions from knowledge that is held only by those who work in the situation. The different perspectives of the various 'partners' in the workplace have to be recognised. The manager needs to know what is to be done, how much it will cost, and if it will work. The people working more directly on the tasks also need to know what is to be done, whether *they* can do it, and what changes it will introduce into *their* working lies. As pointed out earlier, there are more dimensions to the change than those concerned with the interface between work and worker.

Paul Branton was well aware of these matters. His introspective and philosophical nature was allied to an abiding curiosity in how things got done, enabling him to bring to his work in ergonomics a whole range of fresh insights. When we look at the ergonomics scene we recognise that his approach is one of the directions which ergonomics needs, to complement and balance the emphasis – in conventional scientific publications at least – on more detailed research studies. The combination of original researcher and practical implementer of knowledge is a rare one. It was embodied in Paul Branton who has shown us a way which more must follow if ergonomics is to yield its full value.

Note

1. We must be wary of terms such as 'political theory', which imply, by using 'theory', a greater reliability than exists. Such 'theories' are not testable in any real sense, so must be seen more as beliefs.

References

Chapanis, A., 1991, To communicate the human factors message, you have to know what it is and how to communicate it. *Human Factors Society Bulletin,* **34**, 11, 1–4.
Shackel, B., Chidsey, K. D. and Shipley, P., 1969, The assessment of chair comfort, *Ergonomics,* **12**, 2, 269–306.

Chapter 8

From person-centred ergonomics to person-centred ergonomic standards

Friedhelm Nachreiner

Universität Oldenburg, Germany

It is not the intention of this chapter to contribute to the presentation of the person-centred approach to ergonomics which is connected with Paul Branton's name. This will be done in other contributions in this volume and has been done by Paul Branton (1987) himself in his paper 'In praise of ergonomics – a personal perspective'. The person-centred approach to ergonomics should need no further comment from my side since it will be accessible to all those who want to get acquainted with it.

This chapter deals with the other, the practical, side of the coin, the application of a person-centred approach to ergonomics – a science that Branton (1987) called a science with obvious practical significance because it 'helps to answer questions of what should be done' (p. 2). Here again two cases can be made, first *applying ergonomics* in a person-centred approach oneself, and second *getting and helping other people to apply ergonomics*, and to do it in a person-centred perspective. Again, no attempt will be made here to highlight Paul Branton's practical applications of his approach to ergonomics, and his concern for the persons for whom he wanted to provide help in a very direct way. The concern of this chapter is Branton's endeavour for a large scale application of a person-centred approach to ergonomics by other people – *in a person-centred way*.

It is one thing to have a perspective within a scientific discipline as a set of laws, rules and facts, and it is another one to apply this knowledge to problems one is confronted with in one's reality. But even this will produce limited effects only. One of the central concerns for such a 'practical' science in the above-mentioned Brantonian sense, thus, must be the dissemination of its knowledge and 'rules' to those who are directly concerned with or responsible for the solution of the problems that ergonomists care for, man and his work. Therefore, it is only a logical consequence to broaden the scope and field of application of ergonomics in order to enhance the benefits of applying ergonomic knowledge.

There are different ways to do this. One of them is to formulate the rules and the concepts that should be applied. And they should be formulated in a way to make them readily understandable and applicable not only for ergonomists but also for the other people concerned with the design of work and equipment. Even more effective – in the sense of getting people to apply ergonomics – would be to formulate the rules in the form of prescriptions that should or have to be applied, either by legal regulations, by collective bargaining or by agreement, and can be tested for compliance.

One of the ways to achieve this is to produce standards in which ergonomic concepts, guidelines or specifications are laid down. This is why Paul Branton was engaged in the British Standards Institute's committee on ergonomics, the dissemination of ergonomic knowledge on a large scale perspective to those directly concerned with its application. But again, the scope and field of application can be broadened beyond the national level by producing *international* standards, which will be of relevance to more people, especially if they are made mandatory by legal requirements. It seems that this will become true for European standards, which will become mandatory for all members of the EC, thus regulating the application of ergonomics to the design of work and equipment, at least all over Europe and to those who want to sell equipment within Europe. It is this perspective of a widespread application of ergonomics that was responsible for Paul Branton's engagement in the field of European and international standardization in ergonomics.

I remember many discussions with Paul Branton where we talked about how to make ergonomics work. His arguments convinced me that producing international standards would be an efficient way to transfer ergonomics knowledge, which we both agreed to be of practical significance, to those for whom it would be of relevance i.e. designers, managers, workers, and their representatives, as it would improve working conditions for the workers by fitting their job and equipment to the characteristics of man, and at the same time enhance system reliability and efficiency.

I do not remember exactly when I met Paul Branton for the first time, but it must have been in the middle of the seventies, when he made a visit to the Institut für Arbeitsphysiologie at Dortmund, FRG, and I was introduced to him as a senior ergonomist from the UK. Branton was a senior ergonomist who listened to a youngster making his first steps in ergonomics, and who asked wise questions that questioned concepts and perspectives of ergonomics research. This impression was confirmed when we met again on the international standardization scene, Paul Branton as the leader of the UK delegation, and I as a young expert in the German delegation. Discussing formulations for specific statements that would be acceptable to all the parties always involved some reference to the concepts and perspectives to be pursued in writing down a standard for people who might not be familiar with the concepts of ergonomics.

It became clear from Branton's contributions that he was afraid that applying ergonomics in a mechanistic way, e.g. by simply taking isolated prescriptions, for example, on body dimensions, might not give the results and benefits that could be achieved by a thoughtful application of ergonomics, or that this could even become detrimental with respect to the intentions of applying ergonomics. Fitting work and working conditions to man in different but isolated aspects or dimensions, thus, was not the intended outcome. Instead, Branton always tried to point out that work and working conditions should be designed for real persons, having goals, aims, objectives, and behaving purposively in reality.

In contrast with some continental views of ergonomics, which have a more engineering oriented perspective, this was an interesting approach to follow in standardization, because from a psychological point of view, this seemed not only more adequate but inevitable. Such a perspective must lead to an extension of the concept of ergonomics in standardization, and perhaps even further. It cannot be sufficient to regard man as a biological being with anthropometric measures, muscular strength or certain kinds of information processing capacities, but as a person acting purposively under real world conditions. Ergonomic standards would thus have to include these aspects and not just concentrate on the specification of pointer-measurement data.

Clear manifestations of this kind of thinking can be found in some International Standards, in whose development Paul Branton was actively engaged, e.g. ISO 9241 *'Ergonomic Requirements for Office Work with Video Display Terminals* (VDTs)' and here especially in Part 2 *'Task Requirements'* or in the revision of ISO 6385 *'Ergonomic Principles in the Design of Work Systems'*, the basic ergonomic standard, presenting general principles and guidelines for the design of work systems.

ISO 9241 definitely has a person-centred perspective, which is outlined in its Part 1, where the philosophy of the standard is described. Here it is clearly stated that the aim of the standard is to provide for working conditions, including equipment design, that allow for unimpaired human performance at work. It is made clear that physical specification of equipment characteristics is only a means to achieve this end. Hence, the standard specifies physical characteristics as much as possible, in order to help the designer design adequate equipment and the user to use it in an acceptable way, but it does also provide for an alternative approach to achieve the aims of the standard. This can be done by experimentally demonstrating that a certain kind of equipment, for example, a visual display unit or a keyboard, allows for equal or better performance than a comparable unit that complies to the physical specifications in the standard. This, then, is a remarkable step in ergonomic standardization, because it emphasizes the priority of unimpaired performance of the person working with the equipment over the physical, engineering specification, whose function is thus reduced to a 'means to an end' relation. This is quite unusual to engineers or designers, as discussions have shown. But this is a true ergonomic, person-centred perspective: specification is not an end in itself, the central concern is the unimpaired performance of the working person, however accomplished.

Such an approach has many advantages. In the first place it makes clear what ergonomics are about. Second, it relieves the designer from meeting specific characteristics whose aims can equally or even better be achieved in another way, especially when using new or different technologies. It can also account for interactions between conditions, which are generally left out of considerations when specifying isolated physical characteristics in the traditional way. Third, for those who want to adhere to specifications to be on the safe side and avoid any risks, such specifications are given as far as possible.

Besides these advantages the disadvantages of such an approach also become clear: in order to show that a product complies to such a standard by demonstrating equal or even improved performance, where performance is not restricted to the amount, quality or reliability of performance but also has to take comfort in performing into account, an experimental investigation, requiring sound ergonomic knowledge and the necessary facilities has to be conducted, but the facilities to achieve this will not be available to every manufacturer. This, however, can be overcome by engaging ergonomists, who can perform the necessary test, for which the requirements are formulated in the relevant parts of the standard.

It is interesting to note that within the context of the VDT standard, an original technically oriented standard, especially from its German origins, a separate part (Part 2) on task requirements was produced on the international level. Although the working group who later produced this part of the standard was originally intended only to advise the other working groups, dealing with the technical components of the VDT system, with respect to different requirements and to different kinds of usage at different tasks, it very soon became evident that designing adequate equipment would not be sufficient but that normative guidelines on task design, i.e. how this equipment should be used under an ergonomic, person-oriented perspective, would be needed. This part of the standard thus gives some guidance with regard to general principles of task design, where it is made clear that the objective is to ensure efficiency, safety, health and well-being of the worker by providing for tasks and working conditions that avoid overload as well as underload, undue repetitiveness, time pressure or social isolation. Besides these more traditional objectives, task design is required to take into account that man acts as a purposive being when the standard requests that tasks should be accomplished as a meaningful whole, whose significance should be clear to the worker. Tasks should further be designed in a way that uses and develops the skills of the worker and provides for some degree of autonomy in performing the task in order to allow for developing strategies for task accomplishment and goal directed behaviour on the side of the worker, whilst recognising that there is no one best way to do a job for everyone.

It is the '*user-oriented*' approach to task design which asks equipment and system designers to make use of user experience and user expectations when designing and implementing a system that distinguishes this approach from more conventional approaches. It would seem that a lot of elements from Paul Branton's person-centred approach to ergonomics can be found in this part of the standard, in the development of which he took an active part as a member of the working group, and where he defended the approach against more restrictive or conservative positions from other countries or parties. It should be added, that this part of the standard, although not yet officially published at this moment, has received much interest from different parties because of its person-oriented perspective which may indeed lead to a more genuine and beneficial application of ergonomics to system design. Hopefully, this will not only be true in the context of VDT based systems, although this would seem a field with major relevance for future developments in production and work design due to the development and integration of new information processing technologies.

Another place where a person-centred approach most probably will appear in an international standard is the revision of the basic standard ISO 6385 (1981). Although only committee drafts have been circulated within TC 159 'Ergonomics' of the ISO and no predictions can be made about the final results, it can be assumed that some of the ideas that have been brought into the revision process by Paul Branton, the late convenor of the working group responsible for the revision, will be found in the final version. One of these ideas (Branton, 1990) is to call for a validation of system design and to use operators as validators. As a consequence of the person or user-centred approach to system design, the design solution should be validated by having representative workers operate under prospective conditions in a controlled way. Observing real persons performing the appropriate tasks in a controlled or prototype simulation will show whether a design solution is acceptable from an ergonomic point of view, since not only isolated aspects of the solution but also (otherwise unpredictable) interdependencies between design factors can thus be evaluated with regard to their effects or influences on the operator.

It seems that this request for an operator-centred validation under conditions as similar as possible to reality, best characterises not only a – or more precisely Paul Branton's – person-centred approach to ergonomics but also a person-centred approach to applying ergonomics or making other people apply ergonomics in a person-centred way. It would also seem to highlight that 'there is a pressing need for international ergonomic standards in all fields of working as well as living conditions, whatever the development stage of the technologies involved' (Branton, 1990) in order to benefit as much as possible from applying 'practical', 'reality oriented', and, of course, person-centred ergonomics.

References

Branton, P., 1987, In praise of ergonomics – A personal perspective, in Oborne, D. J. (Ed.) *International Reviews of Ergonomics*, **1**, 1–20, London: Taylor & Francis.

Branton, P., 1990, Personal Communication.

ISO 6385, 1981, Ergonomic principles in the design of work systems – General principles. Geneva: ISO.

ISO 9241 (in press) Ergonomic requirements for office work with video display terminals (VDTs) – Part 1: General introduction, Geneva: ISO.

ISO 9241 (in press) Ergonomic requirements for office work with video display terminals (VDTs) – Part 2: Task requirements, Geneva: ISO.

Part III
Significant Brantonian Publications

Chapter 9

Behaviour, body mechanics and discomfort

P. Branton

Ergonomics, 1969, **12**, 316–327
Medical Research Council, London, England

Introduction

Surveying the 20 years since the publication of Åkerblom's (1948) study, one cannot but be struck by the fascination the subjects of sitting and seating exert on ergonomists, designers and the general public. Yet this interest does not appear to have led to much improvement in the general quality of seats, nor has it stemmed the tide of clinical complaints and backpains. The assertion of a causal connection between seats and body malfunctions would therefore not be altogether unjustified, however difficult to prove such a connection may be. Even so, general principles for the improvement of seats still remain to be formulated.

With technological advances and the continuing increase in the time for which people sit, the problem is unlikely to diminish in importance and, even if some may harbour the suspicion that there are no ideal solutions, the search must continue. Perhaps this is the moment to pause and reconsider the conceptual framework of research on sitting and seats. This paper will try to show that behavioural study reveals some gaps to exist in this framework and will suggest ways of bridging them. In our view the gaps result from the fact that the problems are of an interdisciplinary nature and there are three areas for further research in which our understanding could be considerably advanced by joint effort.

The first is the area between body mechanics and behaviour, insofar as it lies between the biological and the behavioural approaches to human action. The second is the area between behaviour and subjective feelings, where we might look for a theoretical basis for research into comfort – and its measurement. In the third area, the technological, demands raised in the other two are to be translated into hardware and then tested systematically to secure validity as a precondition for acceptance by the public at large.

Before discussing each of these research areas, it is necessary to make clear that we shall confine ourselves here to discussing situations of sitting in which comfort and relaxation would normally be the major consideration. Thus we may include not only easy chairs, but also other seats without tables or desks within reach, such as are found in lounges, lecture and concert halls, as well as passenger seating in buses, trains and aircraft. In fact the limiting conditions are that the sitter's limbs perform little or no overt work and that his pelvic complex and spine rest on seat pan and backrest.

Behaviour and body mechanics

Postural variety

Naturalistic studies of sitting (e.g. Branton and Grayson, 1967) have shown that spontaneous behaviour regularly produces a variety of postures with highly significant differences in frequency and duration. These results have since been substantially confirmed by other reseachers using both instantaneous observation and time-lapse films and nearly 45 000 observations are now available for analysis. That sitting postures will, in fact, vary all the time is therefore a basic concept which must be incorporated into our research framework. How can this variety be accounted for? And could we predict which postures will be taken up in a particular seat?

At first sight it might be thought that this variety represented merely random changes of position, an 'urge to move' ('Bewegungsdrang') caused by ischemia as postulated explicitly or implicitly in previous studies of 'fidgeting' (Grandjean *et al.*, 1960, Coermann and Rieck, 1964). Indeed, it is highly likely that ischemia is a contributory factor, in that it creates bodily states which make changes of position desirable. Another likely contributory factor, supporting the idea of such an urge is the considerable pressure on the skin and tissues under the ischial tuberosities (Herzberg, 1955). Ischemia may thus be a necessary condition for that urge to become manifest. But it does not seem to be a sufficient one because it does not explain why on observation certain postures were found to be taken up more frequently than others and held for longer. Neither does 'urge to move' explain why the same person's behaviour should differ so greatly as between seats. Statistical treatment of the behavioural data leaves no doubt that what was observed could not have been postural variation for its own sake and a more complex and specific explanation will be required of how and where ischemia is generated or how it influences spontaneous behaviour.

Another factor usually thought likely to account for the observed postural variations is the possible discrepancy between linear anthropometric and chair dimensions. Here again the relationship may turn out to be more complex than would appear at first sight. It may be remembered that, in the above-cited study of behaviour in train seats, postures differed significantly between tall and short subjects, as well as between types of seat. As the linear dimensions of these two types of seat were almost identical, the differences in behaviour cannot easily be related to misfit between the body dimensions and those of the seats. In this connection the statement by Burandt and Grandjean (1963) – in a somewhat different context – is recalled: 'the exclusive application of an anatomical magnitude for the determination of chairs and tables for office use is unsatisfactory'. Since neither of the usually accepted factors, ischemia and dimensional misfit would account directly for the differences in behaviour, we feel that our observations warrant so far only the limited conclusion that seats do something to the body, or that some kind of interaction takes place between seat and sitter. The nature of this interaction needs to be explored.

Dynamics of sitting postures

The time scale of overt events is very slow and it is therefore not surprising that the mechanics of so-called resting postures have so far attracted little attention. We were first brought to realise the importance of the time factor on viewing the time-lapse films when run 160 times faster than real time. The description of one film in particular will be recalled, as it revealed gradual changes in sequence each of which took 10 to 20 minutes or even longer. These sequences recurred at least 12 to 16 times during a 5-hour period of observation. In each case the sitter slid into a backward slumped posture, propped himself up first with his arms, crossed his knees and then stretched his legs forward, only to end up in a nearly horizontal position. Speaking in terms of interaction, the seat slowly and repeatedly ejected the sitter.

On close analysis the events can be reconstructed as follows. Initially, and because the seat depth was too great for this subject, we assume he had no support in the sacro-lumbar region. Thus, even when sitting upright, his pelvis would rotate backwards over the tuberosities. The upholstery both under and behind him gave little resistance to this movement. If the armrests are used, the trunk is easily raised and suspended by the shoulders rather than resting on the seat. Then, under the backward pressure of the shoulders and with diminished pressure (friction) underneath him, the sitter's pelvis slides forward until a point is reached when the arms can no longer counteract the horizontal component of forces. At this point the angle formed by trunk and thighs is so large that these two levers are driven apart like the unfolding of a claspknife.

To regard trunk and thighs as mechanical levers is, of course, justifiable and not altogether a new concept. Dempster (1955) in a seminal paper refers to the work of Fischer in 1907. Describing the human body as predominantly an open-chain system of links, and applying the kinematic model to sitting, Dempster proceeds as follows.

> The fingers of the hand may be interlocked. . .; the legs may be crossed for seated stability; the arms may be crossed or placed on the hips. In such actions as these, temporary approximations to closed chains are effected. . .Link chains may be connected, as in crossing the knees (*viz.*, pelvis and right and left thighs). . .To the extent that these temporary closed chains approximate a closed triangular, or pyramidal pattern, the less muscles are called upon for stabilizing action at the joints. One may

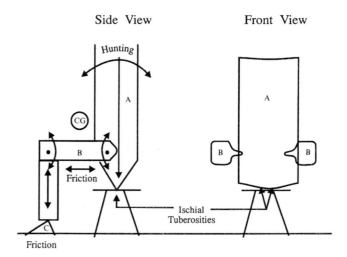

Figure 9.1 The mechanical characteristics of the human body below the waist.

recognize many rest positions involving this principle: crossed arms, hands in pockets, or such sitting positions as crossed knees, ankle on opposite knee, or head in hand. . .As additional joints come into the linkage, the accessory tensions of muscles become more and more important for stability or for directing forces in specific ways. Temporary closed chains that involve extrinsic environmental objects also may be recognized.

Our problem is now to apply such bio-mechanical model to the behaviourally found variety of postures. For if we could in some way measure what a posture does to the body, and what a seat does to a posture, then we would begin to gain control over the design of seats.

The mechanical characteristics of the spine above the waist are known and need no elaboration now. Instead, we wish to consider the structure from the waist down. A schema of this structure and of its mobility can be constructed graphically (Figure 9.1). The point to note in this model is that there are at least four 'degrees of freedom to move', even if the feet are considered to be firmly planted on the floor:

the pelvis can rock over the tuberosities;
the thigh can rotate in relation to the pelvis;
the lower leg can rotate at the knee; and
the lower leg can rotate in relation to the feet.

The notion of freedom to move rests on the consideration that, in the sitting position, the angles at the hip, knee and ankle joints are at about the mid-points of their range of movement and hence in a state of maximum mobility. Any apparent rigidity in this part of the system would therefore depend on muscle action or on external contact with the backrest. These restraints will be discussed later.

If this model reflects the true state of affairs, it draws attention to the mobility of the pelvis and to the flexibility of the link with the lumbar spine at the sacro-lumbar joint. The freedom of the pelvis to move implies not only the possibility of occasional rotation, hinted at when reconstructing the sliding to the slumped posture of the small man in the film. It further necessitates envisaging the possibility of continuous hunting, or relatively fast oscillatory movements of the pelvis rocking over the tuberosities. This possibility arises in all sitting postures in which the top of the sacrum is not resting against the back rest. Indeed, if the lumbar spine, say above L3, were supported while the pelvis remained free to move, such movements would set up a kind of sheering action at the sacro-lumbar joint. This consideration lends weight to the importance Keegan (1953) gives to this joint and the stresses and strains to which is is subjected.

Åkerblom (*op. cit.*), Keegan (*op. cit.*), and Schoberth (1962) have already shown that the angulation of the sacral plate varies with different individuals and Schoberth argues that sitters may rotate their pelvis in an endeavour to bring the plane of the joint to the horizontal 'in order to reduce vertical stress.' This would explain, at least partly, why backward-slumped postures and/or large seat-to-back angles (105° or more) are so often preferred. Such subjective preferences were reported in a number of studies of normal persons (e.g. Barkla, 1964) and of patients with backache complaints (Kretzschmar and Grandjean, 1967).

We arrive thus tentatively at reasons for some of the behavioural observations, in particular of slumped postures. The role of the legs in sitting, and in particular the crossing of knees could also be explained in terms of this model. It will be remembered that the small man at some point in his sliding almost always crossed his knees, and that crossed knees occurred in approximately 30 per cent of all our observations. We believe a fitting explanation for this phenomenon is as follows. It reduces the tendency of the pelvis to rotate. Indeed, when the thighs are adducted and one knee is imposed on the other, a lock is effected across the pubis and rocking is likely to be prevented altogether. Or, if the legs are stretched forward and crossed at the ankles so that the knee joints are locked, the thighs are again adducted and the effect on the pelvis is similar to that of crossed knees. These two ways of counteracting the hypothesized rocking of the pelvis are examples of attempts at stabilization by internal rigidification of the body structure. In accord with Dempster's view, little muscle action would be involved in holding these postures since they approximate triangular patterns. Further considerations in the assessment of these postures are that stretching the legs forward must shift the overall centre of gravity in the same direction, and that at the same time the legs can be used as stanchions to the floor against the forward slide of the body.

Muscle action

Our model leads to the assertion that even in the maintenance of so-called resting postures muscles are involved to a greater extent and degree than seems commonly appreciated. This activity is of the tonic type and its relation to relaxation as an elementary factor in 'comfort' is obvious. It would therefore be of critical importance to measure muscle involvement in various resting postures. This, however, seems to be very difficult. Most published electromyographic studies of sitting have been concerned with work seats and muscle action potentials were mostly demonstrated rather than measured. Moreover, some workers (Lundervold, 1951, Floyd and Silver, 1955, Basmajian, 1962) claim to have detected periods of 'electrical silence' in the sacrospinalis during upright sitting. Yet, in a recent study, Nachemson (1966) by recording simultaneously from the sacrospinalis and the psoas, demonstrated the functional relationship between anterior and posterior spinal muscle groups. In upright sitting, free from backrest, when the sacrospinalis was silent, the psoas was very active, while in leaning forward from the hip the reverse was the case. Evidently one of these muscle groups does the anti-gravity work; which of them it is at a given moment depends upon the position of the centre of gravity at the time.

In the slumped posture, as observed, the spine can be regarded as a flexible open chain of links and, with the expected rocking of the pelvis, if back support is given only under the shoulders, the activity of the antagonistic sets of muscles must be considerable. Since the psoas is accessible only by deep needle electrodes, demonstration of this antagonism is extremely hazardous and it would be fair to say that EMG alone could reveal only a partial picture of the complexity of resting postures.

Attempts to obtain an objective measure of relaxation by more indirect means have been made by the present author. The assumption was that the expected muscle tremor would be shown in fine and fast movements of the centre of gravity and that these might be measured on a force platform. These studies have, however, so far not shown conclusive results. Nevertheless, muscular involvement is a problem still to be faced, since, as Nachemson (*op. cit.*) reminds us, 'The lack of inherent and intrinsic stability of the vertebral column and the importance of trunk muscles are clearly demonstrated if one tries to hold an unconscious person upright.'

Seat features as supports and stabilizers

If it is accepted that the spine and the pelvic complex are loose chain links, it is necessary to consider what support the body structure could derive from normal resting seats. Following on Hertzberg's study (*op. cit.*) Swearingen *et al.* (1962) in an analysis of sitting areas and pressures provide some evidence on this point. They found in 104 subjects that one-half of the body weight is supported by 8 per cent of the seat area (under the ischiae). About one-third (35.2 per cent) of bodyweight is borne by the combination of footrests (18.4 per cent), arm rests (12.4 per cent) and a backrest slightly sloping at 15° rake (4.4 per cent). In effect, the seat pan carries 65 per cent of the weight. Schoberth (*op. cit.*) arrived at similar values by calculations based on Braune and Fischer's work.

To consider the relationship of the seat pan to the structure first. It is the area

Table 9.1 Frequency of occurrence of four groups of postures

Code	Posture	Frequency observed (%) (N = 1644)	Mean duration (min) (N = 10)
1111	Minimal support derived from seat	3.3	2.3
1221, 1222	Full back support from seat and armrests	49.5	15.1
1321, 1322	Slumped, some support from seat and armrests	23.4	11.4
	Other postures	23.8	5.5

of contact with the two ischiae which bears over half the body weight, which means that the sitting body is unstably suspended in respect to the seat pan. This would appear to be the case whether the person is seated on a bare plank or on an upholstered chair, because a soft cushion under the ischiae would impart as little stability and hence rigidity to the system as a flabby muscle or adipose tissue.

This is not to say that, to the sitter, the difference between a hard and a cushioned seat is immaterial. Cushioning does affect the relative distribution of skin pressure over the seat, such that the pressure of body weight would be spread over more than 8 per cent of area. The onset of ischemia is thus likely to be delayed on cushions. If, however, such a cushion were compressed by the ischiae to the point of becoming solid (i.e. if 'bottoming' occurs), that part of the cushion would have the same effect on the system as a hard seat.

As regards the contact of the body with backrests, unless they are raked more than 30° their contribution to support of weight – only about 4 per cent – is astonishingly small. It becomes apparent then that, contrary to expectation and common belief, the effective function of backrests is not to support the body in the sense of bearing weight, but rather that they act as stabilizers.

The extent to which sitters in the behaviour study actually utilized these stabilizers can be gauged from frequency of occurrence of four main groups of postures, and from mean durations as shown in Table 9.1.

This comparison brings out the considerable difference between least supported postures and well supported ones, which tends to strengthen the argument for the operation of stabilization.

To sum up the consideration of body mechanics in relation to behaviour, the sitter is now seen as a dynamically balanced system of open links. Most observed sitting postures are maintained by participation of muscles and the relation a specific posture has to 'comfort' depends, at least partly, on the degree of muscular relaxation it permits. Apart from the possibility of measuring fine and fast oscillatory (or undulatory) movements of the trunk, the various postures may be assessed for their intrinsic stability or rigidity. *In both respects it can be said that at some points in time the seat imparts motions to the body and these tend to be similar for many sitters.*

Behaviour and subjective sitting comfort

Concepts of subjective comfort

The second main area in which conceptual orientation is required if we are to progress towards generalisable statements in sitting and seat research concerns the relation of behaviour to subjective comfort. In the many studies in which subjective sitting comfort has been investigated, very few attempts have been made to define the concept of comfort accurately or to make inter-experiment comparison possible. Some little consolation may be derived from the fact that a similar situation exists in other fields of comfort research. For instance, Teichner (1967) cogently shows this in a discussion of subjective thermal comfort. He concludes that 'All things considered the problem of assessing human subjective thermal responses appears to have suffered from:

1. Lack of use of psychophysical technique, that is methodology which relates behaviour to physical or bodily conditions;
2. Lack of a systematic, theoretical approach to guide research and define validity, and finality, to be discussed below;
3. Failure to consider motivation as a factor in the human response.'

Our general problem is similarly twofold. What is to be measured? and How is this to be done?

At this point, it seems to us essential to bear in mind that the aim of our research is to find out about the comfort of seats rather than the comfort feelings of a person and that the experimenter uses persons as channels of information about seats. Experimentation must therefore not merely aim at obtaining statements about general comfort. It must attempt to point to specific design features and seat characteristics. Moreover, we must consider the possibility that comfort as such is not amenable to measurement and that working definitions of it may have to be modified.

Most investigators assume that the term comfort denotes a feeling or an affective state, which varies subjectively along a continuum from a state of extreme comfort through indifference to a state of extreme discomfort. We find it most difficult to envisage deriving extreme feelings of well-being merely from sitting however good the chair may be. In our view, therefore, the possible continuum extends only from indifference to extreme discomfort. The absence of discomfort denotes a state of no awareness at all of a feeling and does not necessarily entail a positive effect. Similarly, the absence of pain does not necessarily entail the presence of pleasure. This consideration does not merely affect the construction of comfort scales for the eliciting of subjective judgments, to be discussed later. It leads further to the search for criteria in the sitter's behaviour which validly express comfort. It may be that motivation, or the purpose of sitting, provides a criterion. After all, seats are used for a variety of purposes and it is perhaps by their efficiency for these purposes that we may judge them.

Motivation to sit

In the previous section attention was drawn to the need for understanding the mechanical characteristics of given postures and we arrived at the view that postural behaviour may consist of attempts at attaining relative stability of body structure in a physical sense. This implies that in observing postural changes we observe the operation of an underlying purpose and, indeed, our major premise here is that behaviour is not random but purposeful, motivated or expressing needs. Welford (1966) has said 'If we know how motives operate, what they are becomes of secondary importance.' Nevertheless to speculate in psychological terms, the purpose of sitting might seem to be the achievement of maximal comfort. But this could easily overstate the matter, for maximal comfort would be found in sleep. In that case we should more appropriately investigate the efficacy of beds than that of seats. But the human is not to be regarded only as a comfort seeker. He may have a motivation to rest, but if too much rest is given he also seems to seek stimulation from the environment. We would thus not equate maximum bodily comfort with an optimum state of *sitting* comfort. Our case is that we normally sit to some purpose quite unrelated to the shape and properties of the seat and that sitting, like all postural activity, is only a means to another end. In rest seats we sit for a compound of primarily social and personal reasons with the secondary purpose of 'taking the weight off our feet' while listening, conversing, looking at television, or just daydreaming while being transported from A to B. We do not seek comfort for its own sake but rather seek to attain a state which is optimal for the pursuit of these other purposes. Observations of verbal and postural behaviour are then to be interpreted in terms, not of the experience of comfort, but of motivation to avoid interference with primary activities, or of avoiding *dis*comfort.

Comfort achievement is now assigned to a secondary, as it were auxiliary role, which by no means diminishes the importance of research into it, but rather raises it to a psycho-physical problem of attention and discrimination.

This conception may bring sitting comfort research into a field of current psychological experimentation in which performance of two concurrent tasks is measured. A seat may be measurably 'inefficient' to the degree to which it interferes with the primary activity. It also makes experimentation amenable to treatment in terms of signal detection theory (Tanner and Swets, 1954), in that the seat characteristics can be regarded as a source of 'noise' interfering with the 'signals' emanating from other sources competing for the subject's attention. For example, during a lecture we may become aware of disturbing features of our seat to the detriment of our intake of the speaker's words. What we would be trying to measure would be fluctuations in the person's *tolerance of discomfort*.

Problems of measurement

Some problems of measurement of subjective responses remain to be solved. Two of these, set by the limitations of human functions, can be raised here: they concern verbal and memory capacity.

The eliciting of verbal judgements in seat assessment presupposes that something can be verbalized. The verbalization of feelings presents, of course, a longstanding problem in experimental psychology. Moreover, to gain general statements that, say, one seat is uncomfortable or even comparative statements that one seat is less uncomfortable than another are insufficient guidance for the improvement of seating. Judgements should at least be such that practically useful information is gained about either a distinct feature of the seat or a body region, as was done for instance by Grandjean and Burandt (1964).

But, unlike in the areas of environmental comfort research, in seat assessment the subject's attention cannot always be readily focused and related to physical variables as yet not clearly defined. Also, if the subject knew what was wrong with a seat, our work would soon be done. The analysis of postural behaviour presented in the previous section suggests that the difficulty lies in the fact that posture maintenance is a very primitive and deeply ingrained skill, acquired in early childhood. In the brain, this skill, like other motor skills, has very likely an 'enactive representation' rather than a 'symbolic' one (Bruner, 1964). That is to say, like walking or bicycle riding, sitting behaviour is demonstrable but not readily accessible to introspection and verbalization. It may thus be that observation of behaviour is at least as good a guide to comfort as verbal judgement.

It is, of course, not advocated here that we do away with subjective judgements in comfort research. In order to avoid the problem of verbalization, the present author has experimented with the use of a hand dynamometer for the expression of feelings of bodily tension, albeit without success in relating such responses to given seat features. This would nevertheless be one way in which the subject's task of evaluation could be facilitated.

The other measurement problem concerning memory capacity, relates to experimental method and the formulation of questions to the subject. It seems that in experiments in which two or more seats are compared and ranked in any order the fact is overlooked that reliance is placed on kinesthetic memory. The use of rankings or of pair-comparisons of seats to determine their 'comfort-retaining quality' (Slechta et al., 1957) requires subjects to remember what the previously presented seat 'felt like'. Any interval of more than a few minutes between presentations must make these comparisons between feeling states very volatile. With very careful experimental arrangements these methods might produce stable information about a factor of short-term comfort, but judgements of long-term comfort would be very unstable and unreliable.

Of the alternative methods, employing absolute ratings of one seat at a time, an extensive survey of past work found that the most consistent results were obtained from simple unstructured rating scales (Slechta et al., op. cit). Two further aids have been used with some success to focus the subject's attention. One is the use of a body-part manikin (Bennett, 1963) or a body-map (Shackel and Chidsey, 1966). These greatly facilitate subjective reports by avoiding verbal description of the locus of felt discomfort. The other device for eliciting simple but reliable judgement seems to be the use of checklists, e.g. by forced-choice questions like 'The seat is too high/just right/too low'. Though some of them are very laborious, these methods have the double advantage that they may lead to practically useful information about specific seat features as well as allowing more extensive statistical treatment and hence greater refinement of theoretical results. Whatever judgements are elicited, the basic scientific requirement remains to be satisfied, namely that such judgements be set against some objective, physical or behavioural measure. Perhaps here we must take up the challenge Teichner (*op. cit.*) has stated, to devise a 'behavioural preference' method.

Some technological problems of rest seat design

From the foregoing considerations some demands of a technical nature arise for seat design and construction. They concern dimensions as well as other physical properties of the materials to be used in rest seats. The first demand relates to *linear dimensions*. It is unlikely that discomfort can be avoided by simply matching each body dimension with the equivalent seat dimension, as if the interface were static. What seems to be critical is the relation of any one dimension with the others and with expected sitting behaviour. For instance, the relation of seat depth to seat height depends on the purposes of the seat. If a domestic chair is intended for resting postures, a relatively deep seat of 43 cm (17 in.) may be satisfactory for women of the 5th percentile, provided that seat was low enough. In that case, legs could be stretched forward without dangling or causing excessive pressure under the popliteal area. A mass-produced rest chair of the same depth but 42 cm (16 in.) high would be inefficient for the purpose of stretching the legs forward – a posture found frequently. Short persons stretching their legs forward would either lose back stabilization at the sacrum or lose support under the feet. Another instance of an anthropometric demand, mentioned in a previous study, was found in the position of wings in an otherwise very much preferred train seat. These wings were either too low, or too close together, or protruded too far to allow the tall, broad-shouldered person full use of the lower backrest and they thus encouraged slumping.

The second demand presents a mixture of linear and angular dimensional and upholstery problems. It is that the seat's *support and stabilizing functions* are brought within the reach of all likely users. Essentially, the supports/stabilizers are required under or immediately in front of the points on the seat pan where the tuberosities rest. In the back, at least the area in contact with the sacrum is critical. The load of the sitter is bound to distort all cushioning and the location of the critical body areas on the seat profile has, to our knowledge, not been systematically investigated. The specification of unloaded profiles may therefore have to be modified, or at least augmented by specific instructions about the mechanical characteristics of underlying cushioning and springing.

The technical difficulty seems to lie in two demands which appear to be mutually exclusive. On the one hand, cushioning should relieve pressure points and spread the sitter's load over wider areas of the seat pan; on the other, it should provide such support that the sitter's sliding into the slumped posture is counteracted. This counter-action is however to be more a restraining or channelling action than total prevention of posture change. Behaviour seems to show that exertion of the combined strength of hamstring and back muscles is irrepressible. Again, if the seat pan is tilted back so that the included angle becomes less than 95°–100°, while cushioning is very firm, complaints of restriction to body movements and digestive functions may arise. If seat pan and back cushions are soft and thick, support and stabilizing becomes almost non-existent. An effective trunk-thigh angle of about 105° should cater for the observed propensity of sitters to lean back and this can be achieved by shaping and by varying thickness and hardness of cushioning over different parts of the interface with the body. Ideally, the mechanical properties of the seat pan upholstery would therefore be such that it reacts to lateral distortion while being vertically compressible. In modern foam cushioning this is perhaps a technical impossibility.

As regards the demands from the combination of cushioning and seat covering, the provision of a moderate degree of surface roughness would add usefully to the resistance against sliding movements of the body. This matter, mentioned long ago by Åkerblom, has frequently been ignored.

Lastly, another required characteristic of coverings and upholstery is that they should encourage and not restrict the dissipation of perspiration moisture and heat at the interface. Very little research on this topic appears to have been done, but a recent study by Garrow and Wooller (1968) demonstrates the importance of this factor in car driving seats.

In application to particular designs, the solutions adopted to fulfil the above demands would, of course, require experimental confirmation in which both technical experts and ergonomists would need to be involved. Validation and fitting trials (e.g. Jones, 1963) can then be carried out with well established procedures.

> This survey of problems raised by behavioural observation in relation to sitting comfort research attempted to show their possible complexity. It suggested that sitters can provide information about seats not only by verbal judgements but also through other channels, such as performance on concurrent tasks. Observation of sitters may be used in addition to indicate the limits of tolerance to discomfort. Sitting behaviour could be regarded as the operation of a balance between needs for physical stability and for environmental and intrinsic stimulation.
>
> The relationship between a seat and the sitter's discomfort is not thought to be direct because his tolerance may be influenced by an overriding but fluctuating motivation to sit. A more direct relationship appears to exist between a seat and the

sitting posture taken up in it. Thus seats may cause postures and a closer analysis of postural mechanics may add to understanding about the quality of seats.

La confrontation des données comportementales et des sensations de confort lors de la position assise, soulève maint problème complexe dont cette revue de travaux tente de rendre compte.

Il apparaît que le sujet assist peut apporter des éléments d'évaluation des sièges, non seulement au moyen d'un jugement verbal, mais également à traverse d'autres méthodes d'évaluation telles que la prise en compte des performances dans les tâches doubles. L'observation du sujet assis peut servir, en outre, à objectiver les limites de tolérance de l'inconfort. Le comportement de posture assise peut être considéré comme une recherche d'optimisation entre le besoin de stabilité physique et le besoin de stimulations intrinsèques fournies par l'environnement.

La relation entre la conception de siège et l'inconfort qu'il procure n'est sans doute pas directe, puisqu'on supporte l'inconfort d'un siège moyennant une motivation qui, bien que fluctuante, est surtout faite du désir de se trouver assis. Une relation plus directe semble exister entre le type de siège et la posture assise qu'il permet d'adopter. La posture est alors directement dépendante du siège, ce qui permettrait, à partir d'une analyse plus approfondie de la mécanique posturale, de mieux définir les caractéristiques qualitatives d'un siège.

Beobachtungen über das Sitzverhalten werfen Probleme auf, die für Untersuchungen der Sitzbequemlichkeit von Bedeutung sind. Es wird darauf hingewiesen, dass Auskunft über Komfortgefühle des Sitzenden nicht nur aus wörtlichen Urteilen sondern auch auf anderen Wegen bezogen werden kann, z.B. durch Ausübung gleichlaufender Tätigkeiten. Beobachtung Sitzender führt zur Bestimmung der Grenzen der Erträglichkeit von unbequemen Sitzen. Das Sitzverhalten wird als Wechselwirkung zwischen einem Bedürfnis für Stabilität und einem Bedürfnis für Stimulierung betrachtet.

Es scheint keine direkte Beziehung zwischen dem Sitz und dem Komfort des Sitzenden zu bestehen, nachdem seine Toleranz von den Zwecken überwiegend beeinflusst ist, wegen denen er sitzt. Die Beziehung erscheint direkt zwischen einem Sitz und den Sitzhaltungen, die darin eingenommen werden. Sitze bewirken Sitzhaltungen, und Untersuchungen der Haltungsmechanik können zu besserem Verständnis der Sitzqualität verhelfen.

References

Åkerblom, B., 1948, *Standing and Sitting Posture*, Stockholm: A. B. Nordiska Bokhandeln.

Barkla, D. M., 1964, Chair angles, duration of sitting and comfort ratings, *Ergonomics*, **7**, 297–304.

Basmajian, J. V., 1962, *Muscles Alive*, London: Balliere, Tindall & Cox.

Bennett, E., 1963, Product and design evaluation through the multiple forced-choice ranking of subjective feelings, in Bennett, E., Degan, J., and Spiegel, J., *Human Factors in Technology*, New York: McGraw-Hill.

Branton, P., and Grayson, G., 1967, An evaluation of train seats by observation of sitting behaviour, *Ergonomics*, **10**, 35–51.

Bruner, J. S., 1964, The course of cognitive growth, *Amer. Psychologist*, **19**, 1–15.

Burandt, U., and Grandjean, E., 1963, Sitting habits of office employees, *Ergonomics*, **6**, 217–228.

Coermann, R., and Rieck, A., 1964, An improved method for determining the degree of sitting comfort (in German), *Int. Z. angew. Physiol.*, **20**, 376–397.

Dempster, W. T., 1955, The anthropometry of body action, *Ann. N.Y. Acad. Sci.*, **63**, 559–585.

Floyd, W. F., and Silver, P. H. S., 1955, The function of the erectores spinae muscles in certain movements and postures in man, *J. Physiol.*, **129**, 184.

Garrow, C., and Wooller, J., 1968, The use of sheepskin covers on vehicle seats, *Ergonomics* (in press).

Grandjean, E., and Burandt, U., 1964, Die physiologische Gestaltung von Ruhesseln, *Bauen und Wohnen*, pp. 233–236.

Grandjean, E., Jenni, M., and Rhiner, A., 1960, Eine indirekte Methode zur Erfassung des Komfortgefühles beim Sitzem, *Int. Z. angew. Physiol.*, **18**, 101–106.

Hertzberg, H. T. E., 1955, Some contributions of applied physical anthropology to human engineering, *Ann. N.Y. Acad. Sci.*, **63**, 616–629.

Jones, J. C., 1963, Anthropometric data: limitations in use, *Archit. J.*, **137**, 317–325.

Keegan, J. J., 1953, Alterations of the lumbar curve related to posture and seating, *J. Bone and Joint Surg.* **35-A**, 589–603.

Kretzschmar, H., and Grandjean, E., 1967, Comfort requirements of patients with back pain complaints, *Paper given at 3rd I.E.A. Congress, Birmingham.*

Lundervold, A., 1951, Electromyographic investigations during sedentary work, especially typewriting, *Brit. J. phys. Med.*, 32–36.

Nachemson, A., 1966, Electromyographic studies on the vertebral portion of the psoas muscle, with special reference to its stabilizing function of the lumbar spine, *Acta orthop. Scandinav.*, **37**, 177–190.

Schoberth, H., 1962, *Sitzhaltung, Sitzschaden, Sitzmöbel*, Berlin: Springer.

Shackel, B., and Chidsey, K. D., 1966, The need for the experimental approach in the assessment of chair comfort (Abstract), *Ergonomics*, **9**, 340.

Slechta, R. F., Wade, E. A., Carter, W. K., and Forrest, J., 1957, Comparative evaluation of aircraft seating accommodation, *USAF WADC Tech. Rep.*, No. 57–136.

Swearingen, J. J., Wheelwright, C. D., and Garner, J. D., 1962, An analysis of sitting areas and pressures of man, *US Civil Aero-Medical Res. Inst., Oklahoma City*, Rep. No. 62–1.

Tanner, W. P., and Swets, J. A., 1954, A decision-making theory of visual detection, *Psychol. Rev.*, **61**, 401–409.

Teichner, W. H., 1967, The subjective response to the thermal environment, *Human Factors*, **9**, 497–510.

Welford, A. T., 1966, The ergonomic approach to social behaviour, *Ergonomics*, **9**, 357–369.

Chapter 10

Train drivers' attentional states and the design of driving cabins

Paul Branton

Invited Paper to 13th Congress, Union Internationale Des Services Medicaux Des Chemins De Fer (UIMC). Brussels (1970)

Introduction: The ergonomic man/machine systems approach

In their constant endeavour to maintain and improve the safety and efficiency of operations, railway ergonomists and railway medical officers share the problems of fatigue and vigilance of train drivers. It can therefore be taken for granted that the two must always work in closest collaboration. However, without detracting from each other's concern for the human operator with whom both are dealing, a distinction in the form of each one's contribution can be made, and perhaps, the difference in approach can be usefully be highlighted.

To this writer, the prime concern of the medical officers seems to be the protection of the individual's well-being and the cure of malfunctions and illnesses. All their training in diagnostic and therapeutic skills as well as their specialities in pharmacology, toxicology and other medical knowledge are brought to bear primarily on railwaymen as individuals.

By contrast, the ergonomist's interest in individuals is mainly aimed at what these have in common with other railwaymen and how typical each one is as a representative of any larger group of operators. The ergonomist's training is quite specifically directed at exploring the limits of common capacities so as to fit the work and the machines to the largest number of men.

Multi-disciplinary and convergent study

In a sense, the practice of Medicine can nowadays be regarded as a multi-disciplinary endeavour. Ergonomics similarly draws together a number of other disciplines, such as engineering, design, statistics and cybernetics with anatomy, physiology and psychology and other behavioural sciences. It is therefore not merely multi-disciplinary but also a convergent study of human action and the

97

physical and mental effort involved in it. In this respect, ergonomics goes well beyond the usual – and mistaken – concern with applied anthropometry, though of course including it whenever necessary.

Operational and statistical methods

Another distinguishing characteristic of ergonomics is in the methods it uses. These are in the main based on operational and functional analysis rather than historical. One usually asks, 'How does this man operate now?' rather than, 'For what (historical) reason does the man do this?' or 'What is the cause of his action?' The explanatory power of ontogenesis and philogenesis is less frequently evoked in ergonomics than in some other disciplines.

Moreover, the search for commonality and representativeness drives the ergonomist inevitably to the use of statistical probability calculations and the determination of significances of observed events, as distinct from reliance on anecdotal statements. Sometimes in unexplored fields, case studies are however necessary before statistical material can be assembled and reliability of evidence tested.

Variability and errors

The third characteristic in which the ergonomic approach differs from the medical, and indeed from other organisational treatments usually practised in industry, and on the railways in particular, stems from the concept of variability of measured events. Ergonomists accept as axiomatic that all human activity is variable, just as any large series of observations and measurement of living matter displays variability.

Especially in human performance, it is our task to define safe limits and variation of performance exceeding these limits is termed 'error'. It is our further task to predict the probability of occurrence of such errors and to reduce it. The old Latin tag, 'Errare humanum est' is for ergonomists an accepted fact, even if railway organisation, in its justifiable perfectionism would legislate it out of existence with rules and regulations. The insight into this reality has been put as 'A man cannot make less than 1 per cent errors and this approaches the ideal'.

Unintentional errors

In this respect, the ergonomist makes an important working assumption: The man subject to his study is assumed to be of good will and strongly motivated not to make errors of commission or omission.

Anyone familiar with the railwaymen's high sense of duty will know that this assumption is not unreasonable. If errors do nevertheless occur they are regarded as unintentional, perhaps in the legal sense 'due to negligence' but not necessarily 'culpable'. In this light the ergonomists' best contribution to ever safer railway operation is to help in extending the limits of safety by rational and objective criteria based on increasing understanding of human functioning.

Man and machine as one cybernetic system

The understanding of human functioning has been greatly advanced in recent years by application of the theory of dynamic control systems. Cybernetics, accepted into general thinking, is usually associated with the 'feed-back' concept. While this is indeed correct, the important point is that it should be *negative feed-back*, in the sense of *error-correcting* or at least reducing the effect of errors.

The concept of 'error' is the same as mentioned above and it becomes meaningful only if we can specify exactly the *correct* performance of the whole system. Correctness is thus regarded as the *purpose* of the system and errors are defined as deviations from this purpose.

The systems approach to the design of driving cabs can thus be shown as forcing engineers, designers, doctors and ergonomists into collaboration to define very exactly the various roles to be allotted to each part of the whole system, and man's role in particular.

The role of the man in the system

In the attempt to carry control theory into practice, solutions are often sought in terms of mechanical automatic devices and, in principle, by taking control out of the man's hands. This is the general trend in industrial automation, and Dr Gallais has put the problem very succinctly:

> Modern military and industrial systems show two significant tendencies in their development.
> – they are becoming more and more complex;
> – they are becoming more and more automatic.

As a result, in these systems, the human functions which used to be those of a *controller* and *active* operator, tend to change towards those of an *observer* more and more passively facing increasingly automatic equipment.

The present author considers these tendencies deplorable, particularly in the railway context, and he does so not from any backward looking sentimentalism, but for a number of good reasons. The first is efficiency, on which Dr Broadbent, one of the greatest authorities on vigilance research, has summed up the situation as follows:

> By reducing the role of the man to that of a monitor, who is rarely required to act, one can be almost certain that the man will be unfit to act effectively when an emergency does arise.

The second reason is an anticipation of social and industrial relations consequences. The holistic nature of the systems approach brings to our attention the inadequacy of half-measures in allotting roles to men. For instance, if we were to attempt merely to keep the man awake for emergencies by giving him a subsidiary task of no significance to the process of train control we would simply fool him and ourselves. It will certainly take the man only a short time, perhaps a matter of minutes, to find out that his role is subordinate. If so, the consequences can be considerable. At first there is a likely loss in motivation, and it cannot be stressed enough that all performance of worktasks inextricably involves motivation even in its simplest and most short-term meaning. Secondary and long-term consequences are loss of skill, loss of status, loss of self-respect and ultimately loss of dignity.

While these are the generally observed phenomena resulting in most modern industrial situations from reducing a man to being a passive machine minder, Railway Operators have so far been spared much of them. It is perhaps because of the intrinsic interest and enthusiasm engendered by train control and its public responsibilities. A serious thought is that such losses of social content in the driver's work affect operational efficiency and could do more damage than possible gains accruing from automation in train handling.

It remains to state unequivocally right at the outset that the man should be deliberately built into the system, and given not only a clear controlling function but also all the technical help and information he requires to control his train safely and efficiently.

Description of the train driving situation and task in psycho-physiological terms

From preliminary investigations and by drawing on the experience of many drivers, operating and medical officers, the situation of the drivers can best be described as one in which environmental monotony is balanced by a state of inner stress. It is asserted now that if such a balance is struck at a high level it may lead to the condition of fatigue, described so extensively at your Vienna Congress of 1965. The next step is to describe in greater detail the monotony and the likely source of stress.

Monotonous environment – physiological adaptation and stimulus inhibition

Physiologists, since Sherrington, Granit and Von Berkesy, know that the nervous system most certainly adapts and habituates quite rapidly to any prolonged and unvaried stimulation, leading to cortical inhibition of stimuli and to extinction of responses. This is so in the laboratory. Why should the same not be true outside the laboratory, in real-life in the driving cab?

However complex the environment outside the cab, it is naturally thoroughly familiar to the driver. For most of the time on passenger trains, while going along a normal scheduled route, the signals will be in his favour and the road clear. His normal response to the sight of a signal is *abstention from action*, so that he does not even stimulate himself by postural-muscular-kinesthetic self activity.

Inside the cab too, certain monotonic qualities of the environment can be found. The noises are relatively rhythmic, and so are the vibrations and oscillations. Even the climatic changes in the cab are in slow and often imperceptible steps.

It is suggested that all this sets the scene for sensory adaptation and central nervous inhibition to becoming operative.

Dynamic reaction to stress

It is further suggested that the relative monotony of the situation alone can be inherently stressful to the driver, because it conflicts with the need for alertness in his consciousness of the responsibility to control the train, especially at high speeds.

Another possible source of stress is his relative isolation in the cab, more particularly the very limited amount of information communicated by lineside signals and instruments. Much of this stress can be ascribed to something like a state of uncertainty in which he carries out his task.

We know, however, relatively little about the effects of stress on perceptual and motor functions. Nevertheless we can say that in the real situation in the cab the driver is surely aware of being hurled along the track at over 160 km/h at the front of some hundreds of tons of train over which he has only a very limited degree of direct control. That he is under stress is borne out by the findings of UIMC members as reported in Vienna in 1965.

It is desirable now to expand on the model of a balance between monotony and stress by looking not only at the environment but also on the driving task to see whether it is both stressful and monotonous.

Analysis of the driving task

At BR we are in the process of carrying out a systematic study of the skills involved in driving and can only give a preliminary description at this stage. What can be said already is that this turns out to be a good deal more complex than anticipated or understood by locomotive builders and designers. The scope for design improvements to be suggested by ergonomics is therefore quite considerable.

The first stage of analysis requires identifying at least two kinds of separate elements in the task, the continuous or 'tracking' aspect and the discontinuous or 'monitoring'. Within each of these such elements as of memory demands, both short-term and long-term, of anticipation, and of smoothness of execution of movements can be detected. Each will determine what might be done to help the driver.

Tracking task

For instance, it can be said that there is an element of continuous tracking in the driver's constant endeavour to keep going at the maximum permissible speed. In its simplest manifestation this can be shown as his attempts to keep the needle of his speedometer steady at a point on the dial in the face of gradients, curves and considering the load of the train among the factors. He is in fact continuously solving the time distance equation in his head, even if he does not know it consciously. He can solve it, moreover, only if he first knows his present position as well as his distance from destination. This complicated calculation is, of course, also carried out by motorists. But they are not as closely tied to time-tables.

His long-term memory is called upon for instance, to anticipate gradients so that he may put brakes on before reaching the top of a hill in order not to gain too great acceleration on the down grade.

As to smoothness of execution, this author has observed drivers who were 'dining-car conscious', while others were less so and the results can be seen in the sudden or controlled manner in which brake applications are made. Just as in any other psycho-motor skill, the powers of anticipation, of finely graded movements made in perfectly timed sequence reveals the virtuoso.

Monitoring task

The other distinct element of skill is more of a perceptual nature: the monitoring of line-side signals. As already hinted at, this frequently involves passive observation of an information source confirming that the road is clear. The relatively small proportion of occasions when suddenly a more restrictive signal aspect is sighted, makes this task comparable to that of a quality inspector at the production line in a factory, and about that quite a lot of work has been done, on the results of which we might well draw.

Again there is a large component of anticipation in signal sighting based on long-term memory of the road. The driver knows quite exactly the position of signals and certain cues in the wider environment of the track would seem to evoke an orienting response in him so that his eyes can be focused already on the right spot by the time the signal comes into his visual field.

The four aspect signalling system in operation on many main lines of BR provides the driver with a further means of anticipation which also allows him to improve this tracking performance by letting him know something which he often cannot actually see – the state of the road at the next-but-one signal.

An understanding of the role which fore-knowledge of events can play in punctuality and in track utilisation, presents a challenge to railway technologists to present to the driver anticipatory information as well as unburdening his short-term memory by display of the state of signals just passed. It will become clear that the provision of such information, provided it is carefully arranged, can help to reduce stress.

As to monotony of the task itself, some vigilance devices and 'deadmen's pedals' can be commented on. The rhythmic regularity with which most of them operate, almost certainly ensures that the driver's physiological mechanisms will adapt to them, thus virtually defeating the object.

There can also be such a thing as postural monotony, inimical to alertness, and pedals which require a man to remain for long in the same posture may equally have effects which run counter to original intentions.

Cab design for alertness

We all know of some instances when a driver has become drowsy or has fallen asleep on duty. (One case has recently become public knowledge through the report of a train crash at Carlisle. One remarkable point which came to light was that the ventilation opening in the roof of the cab had been blocked with rags so as to keep up the temperature inside, and consequently both driver and secondman dozed off.) In fact, though it is difficult to find exact evidence, the periods of very diminished attention are likely to be mostly extremely brief, only a few seconds duration. Our most important problem is therefore what is known as 'mini-sleep'.

What can be done by design and construction of the driver's work place to reduce the chance of mini-sleep occurring? It is obvious that no drastic changes and improvements can be offered or that their effects would be measurable. Yet we feel we must try to apply an integrated systematic attack.

The defense against monotony lies in the provision of carefully graded changes in stimulation. It is well established by experiments that even relatively small changes in frequency, duration and intensity of any one stimulus will re-activate perceptual system once it is adapted. This fact can help us considerably in design because it gives three degrees of freedom to vary stimuli.

One further lesson from recent experimentation concerns the probability of stimulus or signal events. Work on Signal Detection and Decision Theory shows that the human operator is very good at detecting, quite without awareness, the probability ratios of events occurring repeatedly. Once the pattern or regularity in a sequence has become familiar, this information becomes redundant, sensory adaptation sets in and eventually these events will be virtually ignored by the organism at an involuntary level.

In principle it follows that regular, rhythmical changes should be avoided in design of equipment for alertness and that we should build wherever possible into each parameter of the environment irregular or random variability at least to a moderate degree.

Design from the man out

The very first step at BR has been to evolve a method of co-operation and consultation between the operating department, the engineering designers of new locomotives, the industrial designers, the medical department and the ergonomists. In this process, one concept has been found to be helpful to all concerned: 'Design from the Man Out'. By first studying the dimensional and other requirements of the envelope immediately surrounding the driver, we arrive at a specification of minimal demands of the work place which can then be fitted flexibly to different locomotives at the design stage. This can best be demonstrated to all concerned by means of full-scale three dimensional mock-ups of wood, cardboard and light plastic. Although these tend to look most unimpressive, they do allow visualisation and even some trials of accommodating various sizes of people and of equipment to be fitted into the available space. Admittedly, we cannot yet show a finished product, but the development is under way.

To this process we continually try to apply the principles of human psycho-physiological functions set out above.

Cab climate design

Theoretical standard values for air temperature, relative humidity and air movement have been given by Bedford and other workers for fairly fixed environments. To provide within certain limits a randomly varying climate environment is taking the matter one step further towards satisfying human need for variability. Indeed, the present author is convinced that climatic monotony is the largest contributing factor to drowsiness, inattention and accident causation in train driving.

At BR as one of the first steps, we are experimentally installing on currently used locomotives a type of 'punkah louvre' to give the driver a cool air face bath on demand. On freight trains with loose-coupled wagons driven at night at less than 70 km/h then men not surprisingly go drowsy and a supply of cool air to their faces may help to alleviate the worst situations of incipient mini-sleep.

In new cab design, a series of balances must be struck between the generation of convected heat inside and the radiant heat exchange through windows and wall on the one hand, and the bodily needs of the driver for a cool head and warm feet on the other.

We are experimenting with arrangements of ducting, the positioning of thermostatic switches and their tolerances. It is not so much the number of air changes per unit time in the whole of the cab which now interests us but that the heat accumulated under the ceiling soon and imperceptibly reaches the man's head, thus creating the undesirable temperature gradient.

Noise

Noise in Driving Cabs is likely to be more of an industrial medical problem of protecting the individual than an ergonomic one. From the point of view of design for alertness, adaptation to rhythmic acoustic stimulation is very rapid and de-arousing. But a certain moderate level of random noise in the cab can have an arousing effect and reduces errors in a number of specific tasks. It is thus not necessary to attempt at great cost to suppress all frequencies in the spectrum, but mainly the high-pitched noises which are both damaging and likely to be fatiguing.

Body support and working postures

There is considerable evidence that the character of a seat closely affects the postures of the sitter and that posture in turn is closely related to the person's state of arousal. This is because the level of muscular tonus is raised by internal bodily instability. While, therefore, the driver's body should be reasonably well supported against the rhythmic oscillations experienced in riding, it should not be forced into a seat which allows little change of posture.

At BR we are now experimenting with a net seat which will satisfy this demand as well as a number of others. It is hoped that it will not only give anatomically correct support in a variety of attitudes, but will also help to dissipate the heat inevitably generated by the sitter's body. The need for this can easily be observed and drivers frequently complain about it.

Needless to say, the seat must have the correct anthropometric dimensions. What is more, it must be realised that the driver's body forms in effect a link between the seat and the controls he is to operate. The relationships between position of seat and control levers as well as switches and buttons on the console is therefore critical. These relationships, again, are best assessed in full-scale mock ups before the engineering design stage.

Although opinions are divided about whether the man should be permitted to stand while driving, it can in general be postulated that he should be able to move his body as much as possible to avoid the occurrence of ischemia.

Controls and instrument displays

It would seem that the railways have, so far, ignored the large amount of highly successful work by the aviation industry on this particular subject. In the eyes of a systems-oriented ergonomist, displays and controls cannot be treated separately from each other, but are integral parts of the man-machine interface across which the man continually processes information.

In the train driving context of monotony and vigilance, a few general points can be made. On the perceptual side, determined attempts should be made to display, not just state-at-the-moment information such as present speed, but also clear indications of any changes occurring. As has been said above, preliminary study shows that the driver attempts to anticipate conditions ahead of him and knowledge of changes from moment to moment, whether it is speed, power supply, tractive effort or brake condition enables him to know early when to act on the controls.

At present he derives this knowledge from memory, from unspecified cues of changes in the speed with which the visual field streams past him, from changes in the rhythms of various noises and most acutely from 'the seat of his pants' – i.e. from kinesthetic cues.

On the motor side of control operation, two particular points are worth mentioning: placement and knowledge of results. The positioning and the mode of operation of control levers should be such that the driver can have scope for applying what skills he may have to the fullest. That is to say, power and brake applications are better made with infinitely variable than with step function controls; better with levers that are of a size commensurate with the effect on the train, that have larger travel distances than small and very fine adjustments. Certainly, mere on/off buttons in front of the most sophisticated technical equipment are likely to impoverish the man's skills.

At BR we are trying to arrange brake and power controls which operate upright around a horizontal axis rather than those at present in use which are horizontally turning around a vertical axis. Such levers must, however, be at the right distance and height for 'natural' posture to allow smoothness of operation, otherwise the whole purpose is again defeated.

Knowledge of results concerns the control display relationship and more particularly the time lag between a control action by the driver and the arrival of an indication of its effect on the train. At present this form of feed-back is very often much too delayed to be of other than quasi-historical use to the man. If we are to give the man a proper control function, he must have what is known in system engineering as 'display quickening'.

Vigilance and 'Deadmen's Driver Safety Devices'

It will by now be clear that we are questioning the design philosophy underlying some of the currently used devices. It seems to us that they are uneasy compromises, and that they have two basically different purposes. The safety devices really want to detect when the man is suddenly incapacitated or dead. This would be an all-or-none state requiring a yes/no answer. By contrast, we hope it is clear by now that alertness is an continuum of degrees of arousal, and that a man can press a pedal at regular intervals with his feet and yet not be really alert enough to perceive changes in his environment and act on them.

The whole problem is obviously not easy to solve. Ideally, a *safety device* should detect when a man's heart has stopped beating for a certain length of time. A *vigilance device*, on the other hand, should tell us whether a change has actually been perceived by the man. This could take the form of forcing the man to acknowledge by overt movement certain changes in the display.

We have in mind a combination of electronic capacitance devices, such as are already on the market, with carefully designed use of finger pointing procedures, similar to those used by your Japanese colleagues in driver training. It is just possible that such a combination may solve the dual purposes of safety and vigilance.

Ergonomic check list for drivers

As a first attempt to find out what drivers require of their workplace, we have tried to follow men from the moment they enter the cab and listed as many things as we could see for their consideration. The appended list is not claimed to be complete but should only be regarded as a sample of questions so that these points are not overlooked in design.

Check list for drivers

Answer the following questions by ticking the appropriate square.

Is the workplace sufficiently spacious?
Is the area given to hang your clothes satisfactory?
Is the area given for your satchel satisfactory?
Is the area given for your baggage satisfactory?
Is the seat accessible for you?
Is the seat height satisfactory?
Can you operate the seat sliding lever properly?
Does the seat slide satisfactorily?
Are you properly supported when sitting?
Is the leg room satisfactory?
Do the angled foot-rests give you sufficient room?
Have you sufficient knee clearance?
Have you sufficient toe clearance?
Have you sufficient side clearance to the left?
Have you sufficient side clearance to the right?
Have you sufficient visibility straight ahead?
Have you sufficient forward visibility to the left?
Have you sufficient forward visibility to the right?
Can you see persons on the platform through the side window?
Can you operate the sliding window?
Can you reach the power unit while leaning over in this position?
Can you operate the DSD while in this position?
Can you locate your key satisfactorily?
Is the start switch in a convenient location?
Is the directional lever in a convenient location?
Is the brake dial conveniently placed?
Can you see the dial when you use the brake device for testing?
Is the auto-communication conveniently placed for speaking into?
Is the power handle and DSD conveniently located?
Is the power handle conveniently placed with regard to the speed indicator dial?
Is the power handle comfortable to hold?

Is the power handle comfortable to hold when applying power?
Is the power handle comfortable when decreasing power?
Is the position of the brake handle satisfactory?
Is the brake handle comfortable to hold?
Is the brake handle comfortable when applying pressure?
Is the brake handle comfortable when releasing the brakes?
Can you read the pressure gauge when you apply the brakes?
Is the horn conveniently placed?
Can you operate the horn satisfactorily?
Do you have the reach for the horn?
Are the switches for the AWS, heater, windscreen-wipers and lights placed
    in a convenient position?
Are the arm rests comfortable?

Body movement at speed

Do you use the power handle as a brace to support you?
Do you use the brake lever as a brace?
Does the seat give you sufficient support?
Are the foot rests satisfactory?
Is the brake handle difficult to operate?
Do you get thrown forward?
Can you brace yourself against the movement of the carriage/train?
Do you use the arm rests for this?
When stationary, do you use the auto-communication to speak to the
    guard?
Do you lean out of the side window to see the guard?
Do you use a routine for shutting down all systems?
Do you work right to left?
Is the directional lever easy to operate?
Can you operate the key satisfactorily?
Once out of your seat do you have sufficient space to put your jacket
    on?
Assuming you get off the train onto the track, do you have a hand rail to
    assist you?
If so is it located in a satisfactory position?
Are the steps deep enough?
Are the steps wide enough?
Are the steps too close together?
Are the steps too far apart?
Are the steps satisfactory?
Is the cab door satisfactory?
Is the handle situated in a satisfactory position?
Is the handle comfortable to grip?

Summary

An ergonomic man/machine systems approach to the problems of fatigue and vigilance is contrasted with a medical (clinical and hygiene) approach. Though both are equally valid, emphasis differs. The one protects the individual's well-being and cures malfunctions; the other treats man as one of a larger group of operators, all with certain common capacities, explores the limits of these capacities so as to fit machines to them.

The driving cab represents the contact area, the interface, between man and machine. Across it information is passed back and forth and the nature of this information is discussed. It is shown that certain 'purposes' must be implemented into the system right from the beginning of its conception. For this reason, the role of the man must be considered at the earliest possible moment.

An attempt at understanding the operations and psycho-motor skills will be described, for the purpose of allocating to man and machine those functions optimally suited to each.

For immediate, practical application, but without necessarily neglecting basic research, consideration is given to positively enhancing attention, and relieving monotony rather than searching for specific fatigue factors. Examples will be given for design of environment and equipment to enrich climatic conditions and enlarge the skill content of the driver's task in the face of increased automation.

Resume

Le concept d'une cabine de conduite en regard de l'endurance et la vigilance des conducteurs de trains

L'application des principes ergonomiques au système homme/machine se confronte aux données clinques et hygiéniques. Toutes deux ont leur importance. Les unes mettent l'accent sur le bien-être de l'individu et corrigent ses troubles fonctionnels; les autres traitent l'homme comme faisant partie d'une large communauté de travailleurs ayant des capacités communes et tâchent d'adapter les machines à celles-ci.

La cabine constitue le lieu de contact et la zone de liaison entre l'homme et la machine. Ici les informations sont échangées et prêtent à discussion. Dès le début du concept certaines finalités doivent être observées mais le rôle de l'homme prime.

L'auteur essaye d'analyser les activités et les aptitudes psycho-motrices, pour assigner à l'homme et à la machine la part des fonctions qui leur convient le mieux.

Sans abandonner les investigations fondamentales il est préférable comme solution pratique d'augmenter la vigilance et de diminuer la monotonie que de rechercher les facteurs qui sont à la base de la fatigue. Des modèles d'environne-ment et d'équipement sont décrits pour améliorer les conditions climatiques et augmenter l'aisance du conducteur par une automatisation poussée.

Chapter 11

Ergonomic research contributions to design of the passenger environment[1]

P. Branton

in Passenger Environment, Proceedings of a Conference organised by the Institute of Mechanical Engineers, 1972

Abstract Customer-oriented marketing requires understanding of passengers, and of their personal needs and responses to the environment. Evidence about human capacities and actual behaviour is provided by ergonomics research. This paper deals with some problems of collecting ergonomic data and, in particular, of eliciting reliable information in verbal form and turning it into quantitative specifications for designers and engineers. The limits set to the physical characteristics to be built into the carriage environment by the inevitable variability of capacities and activities of passengers, representing a broad spectrum of the whole population, are discussed. Passengers' feelings evoked by functional and aesthetic aspects of environment will be considered.

Introduction

Marketing specialists tell us that the railways should become customer oriented rather than remain product oriented. To achieve such orientation requires the effort to understand the customers' needs so that the environment can be deliberately and rationally designed to fit them. Ergonomics research, as it describes passengers in terms of their physical, physiological, and mental capacities and needs, helps to give added reality to design. As evidence about real behaviour and actual or predictable use of design features is brought together, so design should increasingly meet realities of use. This paper is mainly concerned with methods of collecting ergonomic data and evidence rather than with design solutions.

Reality as seen and felt by the passenger, if fed back continously and systematically, can help the design process. There is, of course, always some feedback from a vocal section of the community but the problem still remains of distinguishing the singular from the generally held view. As far as the human factors of body sizes, shapes, movements and other personal characteristics, 'user' needs, and responses are concerned, the railways are in the same position as a number of modern mass systems, such as the Post Office. These systems

111

cannot select some people as users and reject others, and therefore designers must be prepared to cater for all members of a statistical significance – how far a particular event or expression of opinion represents the population at large – is one of the ergonomist's main contributions to rational design. Representative evidence helps to avoid situations in which rare specific cases or anecdotal evidence are made the basis of decisions on sometimes very costly courses of action. Collecting representative evidence about the behaviour of individuals and then drawing general conclusions presents problems, and to illustrate the quest to represent reality three areas will be discussed.

The first concerns the difference between answers to questions about measurable facts and actual measurement. The second deals with subjective comfort and the difficulties of describing it. The third describes attempts at demonstrating the existence of consistency in the subjective feelings evoked by environmental design. All three try to bring out the complexities involved in asking people questions and the reliance to be placed on the answers. In other words, one considers attempts to turn subjective statements and judgements into objective evidence.

Factual questions and simple measurements

One of the earliest experiences in ergonomics training and practice is the realization that what people say does not necessarily reflect what they will actually do or what is measurably the true state of affairs. Ergonomists accept that this divergence will occur in questionnaires and surveys. This is not the time or place to go into reasons why this is so; that is a matter for psychological theoreticians and philosophers. The ergonomist does not despair but accepts as axiomatic that people answer questions in good faith, honestly, and 'to the best of their knowledge and belief'. He operates on these answers as expressions of belief only and is aware that they are bound to be held more or less firmly at the moment they were given.

It is, for instance, well understood that it is unwise to ask in questionnaires for a person's height or weight. Answers tend to be unreliable. It seems that short people in particular state that they are taller than is actually the case. The present writer admits to honestly believing for some years to be more than one inch taller than his true height. He was therefore not too surprised to hear of the results of a questionnaire administered some years ago at Waterloo Station, which showed the potential customers of a restaurant to be 'on average two inches taller than British males'. A layman might have thought that most of the respondents were guardsmen or business executives – usually chosen as much for appearance as for other characteristics. To the trained ergonomist, however, such a statement merely revealed lack of sophistication in the 'market survey', and made the rest of the survey data just a little suspect.

One conclusion from this example is that one should build into questionnaires some means of verifying the results. Another is the need to extend data collection, wherever possible, well beyond the asking of questions into behavioural observation and measurement. To look at what people actually are or do may be a good deal more difficult, but it enhances the validity of research data considerably.

Much depends also upon the purposes for which data are collected. Market research is usually in search of 'the average passenger' for more general management decisions, while ergonomics has to be more specific in supplying designers with information relevant to their detailed work. Yet, sometimes there is in the raw data from the former further material for use in the latter. Data collected by market researchers on passenger preferences in regard to luggage were recently subjected to further analysis and yielded formulae to predict dimensional ranges of luggage likely to be carried by passengers on certain trains. In the hands of designers these will be useful not only for space requirements in carriages but also in the design of such equipment as trolleys.

Subjective comfort

The relation of personal comfort to environmental design is controversial and the term 'comfort' itself has, so far, eluded adequate definition, let alone measurement. This applies with equal force to its various aspects, whether thermal, acoustic, ride comfort, or seating comfort. The central problem is, in fact, to relate measurements of a physical nature to subjective, and essentially private, experiences of feelings. Experts in each of these environmental fields admit, when pressed, to unease as to the validity in terms of human experience of technical standards set at present. Since a similar situation seems to exist in other contexts – e.g. in the sceptical attitudes of the lay public to some modern buildings with air conditioning and large glass windows – at least some caution about accepting present standards is indicated. To support or reject this statement scientifically is the very essence of ergonomic research.

Carefully controlled studies in which large numbers of people complete questionnaires, such as those used in the study by Barwell and Clarke presented earlier in this Conference, leave little doubt that some agreement can be obtained among passenger subjects in their verbal expressions of discomfort in various vibration conditions. The same has been found in a number of studies of seating. However, the meaning of the results and their applicability to design is not so clear.

As will be shown, questions on comfort may not be directly accessible to meaningful verbal description, and in asking him one may expect too much from the passenger. Words may simply fail him.

Comfort versus discomfort

It has been said that it is impossible to define or measure health, but only the lack of it. The same is true of comfort. This contention is based upon a very considerable amount of experimental evidence[1]. It appears that, if at all, one cannot measure positive comfort but only varying degrees of discomfort. This is because the absence of discomfort does not mean the presence of a positive feeling but merely the presence of no feeling at all. There appears to be no continuum of feelings, from maximum pleasure to maximum pain, along which any momentary state of feelings might be placed, but there appears to be a continuum from a point of indifference, or absence of discomfort, to another point of intolerance, or unbearable pain. It would follow from this that the optimal environment is then one of which the person loses all awareness and so can give his undivided attention to whatever other activities he may wish to pursue.

Discomfort, and research into it, turns then into the problem of measuring awareness, as part of the general problem of selective attention to bodily and mental functions. For the present purpose it is not necessary to enter into this area, except to point out that experimental psychologists can no longer regard the human as a mere 'black box' to which stimuli can be simply presented and responses measured. It is now quite clear that vastly more information is received by the sense organs than had been realized, and indeed more is impinging on the organism than the brain can handle without at least some selection.

> There is a subjective factor which supports this view. . . Until I mention it, you are not conscious of the touch of your clothes on your body or the pressure of your chair. Or else it may be that you are conscious of these things, yet not of this speech which has been striking your defenceless ears. It appears in fact that part of your surroundings is being neglected. . . The evidence from experiments is that this part tends to change with prolonged exposure to a situation . . .[2]

An operational model of the passenger

For the practical purposes of hardware design, the next stage in comfort research must be to set up an operational model of the passenger as comfort seeker, or rather as 'discomfort avoider', so that equipment specifications can be tailored to him. An attempt at this author's model follows.

Passengers as persons can be regarded as self-contained entities who carry around a set of dynamic fluctuating regulating systems and memory stores full of their own representations of the world around them. The dynamic systems can be body-functional (physiological), the overt behavioural, and the purposive.

Awareness of any of these levels of function is possible but by no means continuous. Indeed, the empirical evidence fits best if attention is also regarded as fluctuating, although predominantly active on the purposive level.

On the level of internal body function, the system has been fittingly described as homeostatic, or self-stabilizing while in relation to the external physical environment it is self-adaptive. If it is hot, we perspire; if cold, we shiver. If we are tilted, we right ourselves. Fortunately, the natural earthly environment is relatively stable, even if only within a fairly broad band of conditions. It might thus be thought that all is well as long as the stable environment persists. Unfortunately, however, for the environmental engineer, an additional characteristic seems to be built into the human system – its appetite for change and variation. It manifests itself more particularly with people living in modern, technologically advanced environments.

The constant stream of stimuli from outside the body, competing with what goes on inside in the business of keeping alive, upright, and active, is too much for us and incoming information is severely filtered. In particular, if environmental stimulation is cyclical, monotonous, or simply non-random, signals are suppressed by the organism. These phenomena are well known by psychological and medical practitioners as habituation, sensory adaptation, and cortical inhibition. Such suppression occurs somewhere between the surface receptors and the cortex of the brain where events might, so to speak, become accessible to 'consciousness' or awareness, and hence to verbal description.

Only the perceptible and unusually strong changes in the incoming stream are reported 'up the line', though not necessarily brought to clear awareness. However, it is the organism which 'decides' what is unusual and therefore interesting. This decision mechanism would seem to operate on the purposive level and is capable of overriding the other levels of function. Purposive or programmed activity, even if removed from awareness, will set new limits to the self-stabilizing system. Again, these limits will combine personal safety with the strength of the person's purpose in pursuit of his on-going activity. As long as the environment remains safe and adaptation enables him to ignore it, the person's purpose will continue to compete with the bodily demands for change and variation. The organism will seek stimulation and, if no other change is possible, a self-induced increase in sensory input will occur to compensate for otherwise monotonous environmental conditions. The person, in this case the passenger, can thus be regarded as a self-perturbing, rather than adaptive, set of systems. It is this author's conviction that the self-perturbation is amenable to observation as it is likely to manifest itself on the overt behavioural level of function.

This view of man as a self-perturbing, stimulation-seeking being has important implications both for the study of discomfort and for the environmental engineer. It means that researchers asking people 'Are you comfortable?' have to reconsider what they demand from their subjects. If the subjects are in a state of adaptation, with information suppressed at the body's periphery, meaningful answers will be hard to come by. If in the process of self-activation, the subject may be vaguely aware of unease or restlessness. This may be of academic interest but will help little for the better design of the environment.

The formulation of questions becomes all important. Most usefully they should deliberately direct the subject's attention to quite specific aspects and features of the environment or to specific parts of the person's anatomy. This specificity is critical, whether ride quality, air conditioning, or seating is under scrutiny. Otherwise words will again simply fail.

Additional methods

Observation and unobtrusive measurement of natural, unselfconscious behaviour of passengers are not altogether new methods, but still relatively rarely practised. They require development and present problems of recording and quantifying critical events. They are not necessarily a matter of costly instruments but of patience and of knowing what to put down with pencil on paper. Once appropriate records are obtained, behavioural observation becomes a powerful tool and may, almost on commonsense grounds, override some opinions expressed verbally by passengers.

The XP64 evaluation study of train seating[3,4] provides a case in point. It may be remembered that the films taken, especially the one of a short fat man, showed passengers to be sliding out of their seats slowly but inexorably many times during long journeys. Yet the little man in particular firmly maintained that he had been 'comfortable' throughout. Having seen his writhings, one could only conclude that his stated opinion was greatly influenced by two extraneous factors. One was that it was known he had never sat in a first-class seat for any length of time; the other that the softness of the old-type cushions was strongly associated in his mind with 'comfort'.

It is worth mentioning that Dutch railway ergonomists recently applied and refined the technique referred to above. They confirm its findings and value and thus lend support to the basis of BR seating research. Such independent confirmation, leading to co-operation and mutual criticism, is to be greatly welcomed.

The poster of the beautiful young lady announcing this Conference is particularly satisfying to the present author because he is one of the originators of the seat which supports her. Viewers who are able to transfer their attention to the seat – or body suspension system – may note that it incorporates many features which will alleviate discomfort. It is based on principles evolved over 10 years of research into anatomy, physiology, and psychology.

Its main features are:

1. Its form provides comfort for a maximum range of passengers.
2. Tensioning of the netting ensures adequate support without constrictive pressures.
3. It permits freedom for an infinite number of seated positions.
4. It prevents the discomfort that can result from over-soft conventional upholstery or the 'too-hard' effect which also sometimes results.
5. It dissipates body heat.
6. It is light.
7. It is cheap (high cost traditional upholstery is omitted and a cheaply replaceable net substituted).
8. It adapts easily to new standards of interior design.

It is clear that this represents a new concept of seating for passengers, which is highly adaptable to a considerable variety of applications. It represents a combination of the orthopaedic and other health demands for a good body support with the general requirements known to contribute to people's feelings about discomfort and its relief. In other words, it attempts to satisfy what people want with what the doctors say they ought to have – certainly a most difficult task. The ergonomists eagerly and confidently look forward to an opportunity to furnish proof of what, in the spirit of this paper, remains an assertion. Only when a sufficiently large sample of unbiased passengers have sat for appropriate durations, and given their subjective reactions, will decisions about these seats be valid. One hopes that this or similar seats will be in use by the 1980s, if not before.

The behavioural studies of expressions of discomfort are not without problems in interpretation of the findings. But, in contrast to questionnaires demanding answers about a private, inner feeling, they have the merit of evidence which is overt and recordable for objective assessment. Ideally, one would hope to find reasonably close correlations between the two methods. Unfortunately, this has not always been the case. Here again, as in the area on factual questions, a divergence between what people say and what they do is demonstrable. In the case of comfort research, while both sources of information must be taken into account, greater weight given to the objective method would seem to provide a better and more rational basis for design. Opinions consciously expressed by passengers – provided, of course, that sampling was representative and the questions properly formulated – cannot be ignored and present a challenge to the researcher to revise and refine his methods.

In practice, design solutions are often dictated by compromises extraneous to the comfort problem, e.g. cost, material availability, urgency of need for a decision before adequate evidence is at hand. It should be clear by now that the problem of measuring discomfort is not an easy one, as the ergonomics specialist is the first to admit. However, this admission is not to be used as an excuse for neglecting his advice; it merely underlines the need for early and close teamwork across departments and disciplines.

Present 'state of the art'

The above arguments applied to the four areas of comfort research – thermal, postural, noise, and vibration – give rise to a broad statement:

Present standards in each of these areas have been evolved from static concepts of 'comfort'. Consequently, advances in empirical research, together with lay assessments, now show up inadequacies and throw some doubt on the approach underlying design of equipment. A new orientation is gaining ground. Related to thermal environment, the case has been stated clearly by Teichner[5]:

It implies that there is no optimum or ideal fixed set of physiological conditions, but rather that the ideal set of conditions are represented by optimally varying physiological levels.

As an engineering consideration this means that the ideal thermal environment is not a constant arrangement of air temperature, ventilation, relative humidity and radiation, but an *optimally programmed* arrangement of these variables. We assume that the programmed variations are small and limited in range.

The same considerations apply, *mutatis mutandis*, to the other three comfort areas in the passenger environment. The provision of constant conditions may end in an impoverished environment; moderately variable conditions result most likely in an enriched one. This may not amount to an immediate answer to the engineers' demands for exact specifications, but it directs the attention of research and development teams to a somewhat different approach, perhaps leading to different and less costly design solutions.

Subjective feelings and interior design

On the choice of pleasing colours and shapes for carriage interiors, designers face a problem similar to the one mentioned above in regard to comfort. They realize that they cannot please everybody all the time and, if they wish to please others as well as themselves, they may be interested in some recent studies attempting to find out *how many* passengers might be suited some of the time. Two studies of passenger responses were carried out in which subjective verbal judgements were elicited from large numbers of travellers (26 000 judgements from 1500 passengers in one case, 68 000 judgements from 280 people in the other). Both used similar methods – 'Semantic Differential' scales and multivariate statistical analysis – applied to different aspects of the carriage environment.

Instead of searching for the chimerical average passenger, these set out to estimate the range of expressed feelings and to establish cumulative percentage curves of the frequency of occurrence. Their major value has been to demonstrate that certain regularities (or perhaps population stereotypes) do exist and that it is possible to predict broadly what feelings will be evoked by given specific colour schemes, space arrangements and other functional design features.

Functional acceptability

One of the studies concerned itself with the general acceptability of various carriages and features in terms of such opposing attributes as

noisy	quiet
relaxing	tiring
cramped	spacious
friendly	impersonal
communal	private

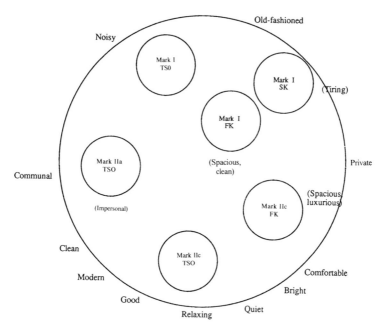

Figure 11.1. A 'map' showing each carriage type in relation to the dimensions of the semantic differentials.

Statistically significant differences between judgements on different types of existing coaches were found and a graphic representation of these is shown in Figure 11.1.

In addition, 18 suggested improvements were ranked for preference. Again, the relative importance of these expressed needs differed significantly between carriage types. The overall order of ranks is shown in Figure 11.2.

Two sets of findings demonstrate the usefulness and sensitivity of this method of collecting data. One concerns the changes in response with level of occupancy as noted at the time of administering the questionnaire, viz. 25, 50, 75, and 100 per cent. In certain types of carriage an optimum level of 'comfort' seems to be felt at around 75 per cent occupancy. At this level the carriage was perceived as more communal, more friendly, more modern, and more comfortable than at other levels of occupancy.

There may be a moral in this as to the degree to which load factors and design considerations may conflict.

The other interesting observation relates to changes over the duration of journeys in the importance given to certain suggested improvements. For instance, 'less vibration and noise', 'cleaner carriages', and 'improved ventilation and heating' mattered more in the first hour than after three hours. 'More space', 'better seats' and 'public address system' rose in importance as the journey progressed.

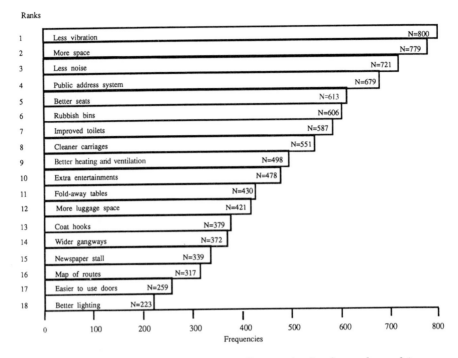

Figure 11.2. Preferred improvements (frequencies for the total sample)

Perceived colour and space relationship

In another study the colour characteristics of carriage interiors were related to spaciousness and to the way in which passengers perceived the character of a carriage beyond its functional acceptability. Evidence from other, well-authenticated studies shows that an important basic characteristic of any colour scheme is whether it evokes receptive or repelling emotions. (Receptiveness itself is not necessarily desirable in an interior, as in some contexts it may be desirable to stimulate movement and activity rather than tranquility.)

The semantic differential scale consisted of 20 pairs of adjectives, seven of which were directed specifically to colour responses while the others were used to establish the desired mood and character in question:

Serviceable	Unserviceable
Dull	Bright
Private	Public
Spacious	Restricting
Modern	Traditional
Unpleasant	Pleasant
Colourful	Drab
Relaxing	Exciting
Luxurious	Adequate

Tatty	Smart
Fresh	Stale
Well designed	Badly designed
Minimal	Sufficient
Subdued	Bright
Warm	Cold
Plain	Ornate
Harmonious	Unharmonious
Hard	Soft
Pleasant	Unpleasant
Bold	Weak

In the present context of carriage evaluation, the first step in analysing passengers' ratings of adjectives is to verify statistically whether or not substantial agreement exists among them. In this case, most of the distributions of judgements showed highly significant agreement for the interiors judged.

The interest of the designer does not, of course, lie only in the ratings themselves but in finding the main attitudinal dimensions and overall impressions on the passenger. In other words, it is desirable to establish not only the desired colour combinations but also their relationship to other attitudes the designer may wish to engender. This is done by finding out which scale ratings tend to go together, e.g. whether a colour rated as pleasant is also rated as relaxing. The outcome of further statistical treatment of the data is a correlation matrix showing, for instance, that 'relaxingness' is highly correlated with 'spaciousness', 'colourfulness', 'freshness', and 'goodness' of design. The general importance to the passenger of these attributes is then brought into relation with the real and specific carriages in the form of weighting scores, by means of factor analysis of the correlation matrix. Such scores tell, for instance, that a certain Pullman interior may be rated exceptionally 'relaxing/spacious', while a suburban open second-class carriage would rate the exact opposite.

It will be quite obvious from a look at the adjectives presented to passengers that the researcher does in no way prescribe to the designer any specific colour, and it would be quite inappropriate to expect him to do so. What can, however, be legitimately demonstrated from this study is the existence of a consensus of moods evoked by a given scheme and arrangement. Having related the character of carriages to the weight given by passengers to certain more specific features, it becomes possible to test next whether the predominance of a particular colour will change the basic attitude pattern.

The purpose of describing this method of eliciting subjective judgements is to show that it is possible to find out how large numbers of people feel about the carriage environment and that testable information can be obtained. Even if there were no objective truth underlying such research, it certainly demonstrates 'inter-subjective' agreements and thus establishes some basis from which customer-oriented design decisions can be made and verified.

The intention of this paper is not so much to present cut-and-dried answers to problems in environmental design, rather it is to show that methods exist by which representative evidence to approximate the passengers' reality can be found.

Acknowledgement

This paper is published by permission of the British Railways Board, but the opinions expresssed are those of the author. The author is grateful to his colleagues at British Rail, to Dr P. Shipley and Mr and Mrs A. C. West of Birkbeck College, London, as well as to Dr J. H. Hogg for information, advice, and comments. Figures 11.1 and 11.2 have been produced by permission of Mr A. C. West, Birkbeck College.

References

1. Grandjean, E. (ed.), 1969, *Sitting Posture*, London: Taylor & Francis.
2. Broadbent, D., 1953, 'Neglect of the surroundings in relation to fatigue decrements in output', in *Fatigue*, London: Lewis & Co.
3. Branton, P. and Grayson, G., 1967, 'An evaluation of train seats by observation of sitting behaviour', *Ergonomics*, **10**, 35.
4. Branton, P., 1969, 'Behaviour, body mechanics and discomfort', in *Sitting Posture*, 202, London: Taylor & Francis.
5. Teichner, W. H., 1967, 'The subjective response to the thermal environment', *Human Factors*, **9**, 497.

Note

1. Paper reprinted by permission of the Council of the Institution of Mechanical Engineers.

Chapter 12

On the process of abstraction[1]

Paul Branton

October 1977

In this paper I shall first describe what is often regarded as the course of abstraction. I shall use an example from Karl Pearson's *Grammar of Science* and then discuss the representation of 'concepts' in mind/brain. Concepts, I shall argue, are percepts idealised in the process of internal representation. This will lead me to try and specify what I mean by the internal representation of an 'ideal' and an elaboration of the mutual relation between ideals and success in skilled acts.

The paper is one of a series planned to stimulate discussion, to lift psychological theories out of their eighteenth and nineteenth century rut and bring them to face the radical scientific and philosophical changes of the twentieth century, e.g. the quantum mechanics revision of physics and a latent demand for a systematic rational ethic. A more immediate purpose of re-conceptualising is to enable us to collect fresh observational and experimental evidence for better fitting theories.

The present problem arose in my mind around a central puzzle in accident research: in skilled human performance, errors are difficult to define exactly in terms of perceptual-motor functions. Indeed, in a paper to the Birmingham Ergonomics Group (1976), I said that it appeared logically impossible to specify exactly the limit of error-free performance without having first defined accurately the successful attainment of the target. Paradoxically, it is the essence of a skilled act that it is ever more refinable and perfectable. As with the hand movement of a pianist, timing, positioning and strength on arrival at almost any target can be so accurately graded that they have the open-ended character of an 'ideal', in the sense of expressing the *idea of the operator*, rather than mere repetition of the fixed programme of an automaton. 'Ideal' may be a strange word to use in a scientific context, but I can find no better.

From this I went on to puzzle about the nature of 'internal representations', a term Kelley (1968) uses as a theoretical construct necessary in describing the control of man-machine systems such as space craft. What might be the forms in which data are stored? What are the necessary and sufficient requirements of such successes of action as are actually observed, e.g. when men land on the

123

moon – in previously unknown territory? How can 'ideals' exist in the mind/ brain? One thing was clear to me: whatever the material and quantitative aspects of these representations were, their most likely quality was abstract, in the sense of being general rather than being tied to one specific event or concrete state only. So we go from errors to ideals and on to abstractions. What operations must take place when one abstracts something from a perceptual experience? We observe what is, without doubt, the success of a skilled person; we also know something about the psycho-neuro-physiological equipment he carries. To reconcile this knowledge with the obvious successes, and to understand the workings of that equipment, is a necessary precondition for the training and teaching of skill, whether to the novice or the patient whose skills have broken down. Equally, if it is to help the operator to control rather than be controlled, 'man-machine symbiosis' can only be achieved when we know more about man's functions. We have here one complex of questions: on the nature of idealised targets, or how and what do we abstract from experience? And what, if anything, is already there before experience?

Having left this complex to 'brood' for a while, a similar one arose on a different level. Whilst collecting material for a paper on the skills of research workers, I came across the work of some friends of mine in the form of a Socratic discourse on critical reasoning (Heckmann and Henry, 1975). They inquired into the stages in thinking from (veridical) perception of concrete events to general, categorial statements about causality and substance. One does not need to climb to inordinate heights of metaphysical speculation to appreciate that research workers essentially deal in general statements about causality and that they assert the existence of lawful regularities in the behaviour of inanimate matter as well as of living beings. The ordinary citizen in everyday life exempts himself from justifying his generalisations. When he searches for a key he rarely says, 'The key cannot have disappeared without trace', and is most unlikely to add, 'because there is a law of conservation of matter'. But if he is a cosmologist, studying the first few seconds after the Big Bang at the start of the Universe, he has to be very careful about generalising from present experience to categorial connections between events. Then again, researchers do not lead two lives, one general and one non-general; they do search for lost keys like everyone else. Abstraction just seems to happen. My friends provisionally concluded that:

> Abstraction is the attempt to formulate the rules by which we let ourselves be guided in experience, and to apprehend these rules as generally valid laws of natural events, as they offer themselves to us in the present state of our experiential knowledge.

The formulation takes account of certain problems of theory of knowledge, and of the effects of developments in quantum mechanics on the concepts of causality and substance.

Without giving away too much, I may say now that one of my conclusions on the skills of researchers will be to point to a peculiarly perfectable combination of realism and imagination. In this field at least, excellence is the most realistic yet farthest reaching flight of fancy.

So, to put it simply and crudely, the important step in scientific advance is to *abstract* from the mass of 'given facts' durable, lawful relations between experiences and discard the accidental, the random and the trivial. There is no point in scientists being ashamed to admit that these are philosophical and indeed epistemological concerns, insofar as they inquire into the sources and grounds of all our knowledge. Willy-nilly, researchers must all be meta-physicians. I am not even sure whether we are on meta-*psychological* ground when we deal with the thoughts and knowledge of researchers, as distinct from the objects about which they think and know something. Anyway, even if it is meta-psychology, psychologists cannot afford to ignore it.

The process of abstraction is therefore of interest not just as topic for psycho-biological study, but also justified as striving for philosophical soundness of thinking. If both philosophy and psycho-biology are to become more scientifically secure, their reconciliation must be attempted again and again.

Description of the process: (First approximation)

The instance which set me off on this train of thought comes from Crowther (1970) who, as a teacher, wished to:

> attempt new methods of mathematical instruction. . .trying to get rid of the formal and given aspects of Euclidian geometry. Euclid makes deductions from perfect points, lines and angles, *which do not exist in the environment* . . . [Crowther] started from an idea in Karl Pearson's *Grammar of Science*, that geometrical conceptions, such as points and lines and angles, are *the ultimate product of a process which starts in perceptions of the actual world*. The mind abstracts qualities from the perceptions until pure conceptions only are left. The notion of a point is the result of abstracting colour and the three dimensions of space from solid objects. [He] constructed a series of irregular solids, painted in mixtures of colours, and hung them in a row high on the wall of the study-room. Beneath this [he] hung a parallel row of white tetrahedra, smaller and closer together than the coloured irregular solids; thus in the second row, colour and irregularity of shape had been abstracted. Then a third row of small flat white circles, very close together, was fixed to the wall; this row was beginning to look like a white line made of white points. Below this row, a parallel white line was drawn. Thus the student could see that the fine white line was a late stage in a process of abstraction from the coloured irregular shaped objects found in the natural world. The aim was to remove the mystery of where the perfect points and lines discussed by Euclid come from. (My italics).

Now, sceptics may say that the procedure descibed by Pearson is unrealistic for other than mathematical-geometrical abstraction in the classroom, and this for at least three reasons: they may consider it to be a) rarely used, b) too cumbersome and c) too slow to be more than a special case of perception. I believe to the contrary, that, broadly speaking, we go about abstracting like this continually in all our skilled activities. But this is not necessarily done consciously or cognitively.

Modern experimental research into perception leaves no doubt about its complex, hierarchical nature, its speed and ubiquity (cf. Haber, 1970). I am referring in particular to the recent discoveries by Hubel and Wiesel (1968), by Blakemore (1971) and others, of single nerve cells operating as 'feature analysers' in the mid-brain and cortex. They open up a new interpretation of much older evidence that had not fitted in anywhere. It could mean that each cell or part of the brain extracts (or abstracts?) from the incoming stream of signals those of its own special interest: colour, brightness, straightness, skewness, roughness, etc. The preferences of, say, certain visual neurones to respond to lines in specific orientations, or even to binocular equivalence, can be interpreted as evidence of structures being used in an abstracting process such as I have in mind here. Since not all stimuli impinging on receptors issue immediately in skilled acts, in the normal time-scale of life outside the psychological laboratory, there is enough time for even the most complex abstracting operations to be performed. It is very likely, though, that a great deal more than we are aware of is turned into generalisations (and used to update the experiential store) (Cf. Bartless, 1958, MacKay, 1965, Bach-y-Rita, 1972). This could be a continuous internal reconstruction of our representation of the external world and may be, if we set aside for the moment his historical, ontological superstructure, what Sokolov (1963) meant by a 'neuronal model. . .a polyvalent model of the stimulus in which all or a considerable group of its properties are represented'.

The main point, not to be lost sight of now, is that the line itself, in the above instance given by experience, impinges upon *an organism already structured and prepared to categorise environmental features into attributes and qualities.*

It would seem that one consequence of asserting this is that the number of attribute-categories provided by the system must be sufficient to discriminate between the variety of possible stimuli. Is this true? Must there be as many pigeonholes as there are attributes likely to be encountered if the manifold of natural events is to be explained? Could this be why in the laboratory to distinguish an ellipse from a tilted circle is so hard? Experimental determination of discrimination thresholds has always been difficult to achieve.

It is noteworthy that researchers like Sokolov who can hardly be counted among the idealists, advocate clearly and repeatedly that internal representations must exist, that they are very accurate and yet not necessarily physical or mechanical functions. Observations of skilled behaviour tend to bear this out (Branton, 1977). It is, however, not clear how the considerable changes in sensitivity from one experience to another can be reconciled with this exactitude. Sokolov, and many others who accept his model, overcome this by postulating the 'orienting response' with its assumption of increased sensitivity '*due to*' intensive direction of attention. This shifts the question to another level, but does not answer it.

True, the postulated existence of feature analysers for each attribute of the environmental stimulus does not, of itself, provide the principle by which some rather than other stimuli are selected for abstraction from the continuous stream of signals. But three points may be noted about the postulate. Firstly, it may give greater precision to the possible location of the seam between the 'out-there' and the 'in-here'. Secondly, such an existential assertion need not lead us into the nature/nurture controversy. I would regard this as a barren argument and hope not to be sidetracked into it. Thirdly, and consequently, I wish to leave open the possibility of interpreting abstraction as an autonomous, spontaneous act. So far, the present view appears compatible with the model of endogenous, self-activated, central events described by Pribram (1976). An earlier, equally appropriate description of spontaneity can be found in Hick (1952).

From percept to concept

Where does the assumption of feature analysis lead us? I am indebted to H. Dücker, the doyen of German experimental psychology, for the following thoughts expressed in a delightful interview earlier this year (1977):

> In the old school they spoke of introspection and self-observation, as if I could observe my Self. But we can speak only of our Selves as experiencing something, and those experiences do not hang in the air. This requires a *postulate*: Experiences must have a *Bearer*. This bearer is my Self. My hearing, my seeing, are activities of this Self. If I see a table, I merely perceive a patch of colour and assert it is a table. That is to say, we attribute the features – colour, extent, etc. – to a bearer. Attributes cannot exist without such as bearer, carrier, or subject. Now, I ask, was Comte somehow in error? Does one really observe the 'thing-itself'. No. I only observe its properties or features. . . When I observe features, I must necessarily *postulate* someone or something behind them.[2]

If neither the general thing-itself nor the specific thing-out-there exist in our brains, but only its attributes are stored in them, it is a short – but very important – step to propose that the combination or collective of activated neurones, 'representing' the features observed, *is that thing for us*. Hence, whatever is stored and represented inside us, is already an abstraction.[3]

As speculations go, this is no new thought at all. Indeed, with variations it has been put forward perennially as explanation of deep-rooted experience. It reflects Plato's Theory of Forms. We may appreciate the aptness and simplicity of language of the following quotation when we realise that 'predicate' stands for feature or attribute:

> It is the 'presence' of the form in a thing that makes anything beautiful or whatever else we say it is. The predicate of a proposition is always a form, and a particular sensible thing is nothing else but the common meeting-place of a number of predicates, each of which is an intelligible form. . .

On the other hand none of the forms we predicate of a thing is present in it completely. . . . Apart from these [predicates, the thing] has no independent reality, and if we know all the forms in which anything participates, there is nothing more to know about it. (Burnet, 1914).

It may sound like a simple proposition of syntax, and perhaps it is meant as one: the subjects of sentences have predicates, and vice versa. Yet, I find in the above quotation three phrases that illuminate our puzzles. The 'common meeting-place of . . . predicates' evokes in my mind the problems connected with localisation of functions in the brain. Anyhow, what functions are we dealing with? Maybe we shall have to revise our categories of 'functions'. (It seems that this revision is being started again by Popper and Eccles, 1977).

The second phrase, 'apart from these [predicates, the thing] has no independent reality', helps me to begin to understand how such vast amounts of experience can be stored in so small a volume, how much easier access to the elements must be than if the store consisted of things represented, and how much more flexible the use of storage is if it becomes independent of sequence of absorption. One of my favourite quotes from Bartlett (1932) is about memory overcoming 'the sequential tyranny of past reactions'.

Lastly, that no form predicated of a thing is present in it completely, explains to me the open-endedness of skillful acts and the ideal nature of representations. I shall remark briefly on each of these points.

The collective of feature analysers activated by a specific thing, or the 'common meeting-place' of certain of its predicates, need not necessarily be located spatially close together. There does not seem to be any reason why each contributory cell or element could not be located or grouped according to some, perhaps, as yet unidentified function. It may be sufficient that synaptic connections exist. Otherwise localisations could be too limiting. The probability is very high that combinations are cross-modally connected, such that the round, reddish-yellow thing, visually perceived by the cured formerly blind patient, becomes a conceptual orange as soon as touch adds the specific tactile predicate of texture to the collective (cf. von Senden, 1960, Gregory, 1966).

There are also such simple cases as that of the schoolboy who thought the distant world was always blurred until he had spectacles prescribed. How was he to know different? The acuity of his feature analysers had first to be set to narrower tolerances. There is not much literature to explain in adequate detail this process which seems to me to be 'corticofugal' (Granit, 1962), and hence probably self-activated, spontaneous or endogenous. Von Holst's efference-copy mechanism, and Miller, Galanter and Pribram's TOTE (1960) are the nearest approach I know of.

Nor need the products of abstraction be isomorphic with their real object. Indeed, the 'shape' of the common meeting-place is quite likely to vary from one person to the other, and within the person from time to time. Perhaps we may liken that shape to the caricatures or homunculi used to illustrate localisation on the sensory-motor cortex in some physiological textbooks. Taking the last two points together, the common meeting-place could be styled '*homunculus dispersed*'.

Of course, to conform both with experience and the philosophical demands posed by the 'Other Minds' problem, the variability of our homunculi must not reflect such extreme subjectivity that communication with others is inhibited too greatly. Does mental abnormality and illness lie that way?

That the thing has no reality – in our brain – independent of the characteristics participating in the collective, allows the *flexibility of recognition*, and the search for attributes to complete the process of identification, so difficult to explain otherwise. At the same time, the phenomenal economy of storage by predication becomes more easily explicable. There need be no limit to the number of objects sharing some single attribute. But no two objects can have *all* their attributes in common. That would offend our notions of time and space. Two things in the same place at the same time cannot but be the same thing.

Ideals represented

To argue that internal representations are *idealised* and that it is not the actualout-there which is represented may be more difficult and more complicated. But this is only so because abstraction is so elementary and primitive a process that we practice it well outside ordinary awareness. School teachers, psychologists and sociologists, all who use notions like Intelligence Quotients, are only too familiar with the thought that normalcy has no exemplar in real life. Neither, as everyone knows, does the 'average person' exist. The numerate scientist, particularly the statistician, should have no problem. But if he will say that all mathematical thinking is abstract as a matter of course, he would be only begging the psychological question. Indeed, as I have argued elsewhere ('Train Drivers' 1978), we may be able to understand mental operations better, if we use the quasi-mathematical analogy in modelling mind/brain activity.

Supposing then the collective of neurones firing when presented with the stimulus of, say, a table *are* – as far as it concerns me – a general table. To the extent to which any one of the attributes of tables is not completely present in the thing-out-there, to that extent it falls short of the 'perfect form'. The degree to which the real out-there may be imperfect as compared with the ideal in-here seems a question of *limits of tolerance of inaccuracy*, not only in the use of language but in all skilled acts.

The ideal target in-here may be the very point at the centre of the bull's eye of the dartboard at which I throw that dart, though I can by definition never hit the ideal but only some poor approximation to it out-there. With *practice and effort* I might refine my definition of accuracy. I thus vary my tolerance as the situation demands which fits well with Bartlett's dictum about skill, whether bodily or mental, having 'from the beginning the character of being in touch with demands which come from the outside world' (1958). It is noteworthy that the same variability of tolerances has recently been shown to exist in the physiological feature analyser cell system mentioned earlier. (Cf. Shinkman *et al.*, 1977).

The perceiving organism reports, as it were: 'Out-there now exists a shiny, brown, square, four-legged, inanimate thing, a metre high with horizontal flat top of smooth texture.' In abstraction, if a predicate can at all be detached from its subject, the detachment can only be from a specific location in space and time. In fact, to analyse is to break up the space/time unity of the thing, in this case the table. (We leave aside for the moment that it is just this break-up which allows us to store the percept as an experience.) It may be that we thus disconnect the specificity of the feature from the feature-bearer and thereby assign the bearer to a class of things-bearing-this-feature. Nelson (1972) refers to this process as 'abstraction from the numerical specificity' (e.g. of persons as bearers of interests). As a thought experiment, it is of course permissible and reasonable to dissect a thing in this manner. How far this parallels what happens in nature is another matter. But we note now that the elevation of the thing to a class is the critical stage in the process of abstraction, because this is a potential generalisation and re-combination, even though there may at the time be only one exemplar in the class. To generalise is then to assert the potential that there could be more than one member of that class. It is then this re-combination of features into events and things, whether empirically given or 'imagined', which constitutes the creative aspect of abstraction. I refer in particular to the creative thinking of scientists and the skills of other high-level performers.

Scientists have the additional problem that their thinking appears to create *new* connections between two imaginary events, such as causal connections, and that these events are stored in the idealised form. The puzzle is not so much about the events as about the notion of causality itself. If concepts are idealised percepts, what about abstract concepts? We may perceive with the senses two events but never the connection between them. (The Gestalt ideas and Michotte's experiments do not seem adequate explanations as proximity of time and place, in particular, are brought into question by the revision of physics due to Bohr and others.) I cannot take this matter further now, except by suggesting that the concept of causality might, with advantage, be enlarged to include not only the usual retro-spective concern with antecedents, but also the teleological explanatory mode, described by Taylor (1964), as a possible way to an understanding of autonomously generated 'prospective causes'.

It is essential to my argument that there exist such prospective causes, whether we call them purposes or what else. The point is that they must be represented internally. I think of an internal representation as an actual process event, an operation upon a template, a coded reverberation. Something, once perceived, becomes lasting; a remnant is formed, preserved over time, transcends the moment of its inner creation and decay, whether or not resulting from outside influence. This process is our *own organising power*, our *negentropic* principle.

By 'Transcends' I mean the transformation which, by nature of the process overcomes the limits of temporal and spatial existence of things and events. (I confess, this formulation seems tautological and hence unsatisfactory. This is because 'the rules by which we let ourselves be guided in our experience' (Heckmann and Henry, 1975) have still evaded me. Perhaps all I have been able to do is to narrow down describing what the steps are from perception to incorporation into experience, but this does not give the criterion of choice for entry into the store. Neither does it deal satisfactorily with what I would call the 'ligatures' or notions of connectivity, which allow us to assemble just those cell collectives and not others. The alternative to assembly would be a random jumble of accidental, associationist, non-connected stream of white noise which is not the case!)

As mentioned earlier, one effect of the – to us – naturally occurring break-up of a thing or event into features is a disruption of its space/time unity. Although time's arrow has been shot, the representation preserves in some as yet unaccounted way the *status quo ante*. Most likely a representative 'cell assembly' or collective, or common meeting place, has a time marker attached, as Bartlett rightly insists (1958, p. 184). In our present terms, this means that an internal representation may exist only in time, *not in a topographical point*, but only in a statistically determinable space. The possibility of this manipulation of time may be thought of as the most significant result of abstraction.

The significance lies in the fact that the internalising of externals opens the way to a radical change in explanatory mode from the causal to the teleological, or from retrospective to prospective explanation of an act.

When it comes to the question of what these internal representations are about, I think, they are as much about the future as about the past. I have argued in the case of a control skill of medium complexity (train driving) that these representations may actually be the subjective consequences of the operator's own intended actions. Mackay (1965) put it slightly differently: '. . .the world is internally represented in terms of the pattern of demand (actual or conditional) which it imposes on the perceiver.' If this is so, two corollaries follow; one is related to causality, the other is about the reality of the outside world.

Firstly, on the level of ends or purposes, the active representation of an end state can, I believe, become a 'cause' of a kind, if it gives direction to the skilled act. It represents the end state internally not only without necessarily having any exact counterpart extant outside the person, but also by having valued and decided the direction. Thus Hull's paradox may be resolved. It will be remembered that he held that:

> In its extreme form teleology is the name of the belief that the *terminal* stage of certain environmental-organismic interaction cycles somehow is at the same time one of the antecedents determining conditions which bring the behaviour cycle about.' (Hull, 1934).

This point, remarked on by a number of authors (Boden, 1972, Taylor, 1964) is clearly central to the theoretical positions of many psychologists until the 1960s. Thus Hull's position is quite consistent with the pure behaviourist approach and, generally, with the hypothetico-deductive method of the natural sciences before Bohr and Heisenberg. But as far as the biological and human sciences are concerned, there have always been some issues unresolved and Hull's ironical remarks cannot pass as adequate justification for denying that humans could behave purposively. For, this is what the statement amounts to and it has held sway too long, No doubt, it is not always easy to be sure whether an act was intended and what the intention was, but to deny to psychology the study of purposes and their role in behaviour can only bring the science into contempt.

Now, although I have used 'purpose' and 'goal' interchangeably and with apparent latitude it should be made clear that the ideal, abstract, conceptual representation is in fact very exact. The nearest analogy in mathematical thinking is the numerical representation of Pi, with its infinity of decimal digits. It should be equally clear that the concept of goal itself is capable of quantitative expression in the appropriate context, say in applications of control theory. Thus Baum and Drury (1976) describe models of the organisation of behaviour as essentially goal-directed when 'routines are selected as a function of their output rather than what they do. . .' If we substitute 'outcome' for output it becomes clear that the *consequences* of an act can be the *organising principle* of it.

Success in the real world

To advocate the existence of an inner world of such complexity, exactitude and time-defeating character as I have done here is to go to the limits of scientific credibility. It is epistemologically impossible to prove the truth of the assertions and all one can say is that they are necessary to explain certain phenomena. Perhaps one line of argument offers proof: In the real world, out-there, we do succeed in the vast majority of our acts. How could we succeed, if all the internal world consists only of unreal, indeed of ideal representations? Resolution of this puzzle rests in the successful achievement of intentions. External, overt, public success is as necessary as the existence of internal ideals to explain the facts of skilled behaviours from the simplest to the highest complex mental ones. To reconcile a large set of unrelated facts, the best fitting explanation is to postulate that a continuous 'trade' is carried on between in-here and out-there.

Taking most elementary acts like standing, walking, grasping, we remember that the stage of perception, mediation and motor action are actually insepara- bly intertwined and only treated apart for convenience of discourse and teaching. It then becomes a commonplace to say that there always exists some ongoing commerce between receptors, memory and effectors. The most striking common aspect of the physiological and psychological mechanisms suggested by von Holst and by Miller *et al.* (mentioned above), is the continuous testing of the world out-there against the internally stored model. By this means the gener- alisations of abstraction are used incessantly to update the experiential store in the normal course of events.

Conclusions – unfinished business

I must conclude by admitting that I am rather short of immediate suggestions for researching into the consequences of what was postulated above. Since most of the dissatisfaction with theory stems from experience in the applied field, one would wish to be practical and take part-problems into the laboratory for testing. But this may yet come.

Of no less importance – to my mind – is the task of securing sufficiently strong bridges to philosophy, especially as regards the two sides of teleology, methodology of science and the other, almost completely uncharted area of ethical or moral behaviour and valuation. Almost every single act we perform is likely to affect the interests of some other person. It is with these problems in view that I insisted earlier on to leave the option of an autonomous person open. Perhaps it will become possible at some time to operationalise ethical acts. We speak easily and glibly of social skills nowadays. When all is said and done, the ideals underlying these skills are moral, and if all our acts can be means to abstract ends, psychologists can no longer sit on the side lines.

Notes

1. This is an unpublished manuscript by Branton. A bibliography is not available.
2. The German expression '*Das Ding-an-sich*' should be translated as 'the thing-as-it-really-is'. However, this is too clumsy for the present purpose and so I use the 'thing-itself'.
3. If we now stipulate that it does not matter in the least whether that thing exists regardless of anyone ever even potentially perceiving it, I believe we need not lay ourselves open to the kind of idealism proclaimed by Berkeley and criticised by Hume. Dr. Johnson would not have stubbed his toe to refute us.

Chapter 13

Investigations into the skills of train-driving[1]

P. Branton

Ergonomics, 1979, **22**, *155–64*

Abstract: Train driving as a control task is one-dimensional, yet very complex. This combination highlights certain problems in our understanding of skilled action. Investigation involving behavioural observations, plus interviews with over 200 drivers and inspectors, showed that the drivers utilise more information from outside the cab than is usually thought. The relevant variables were identified. The limitations to the driver's possible knowledge of the changing state of the system ahead of him lead the study to the goal-directed, purposive nature of his skill. What exactly does he have to carry in his head to achieve the observed successes in time-keeping and safety? Consideration is given to the form of internal representations of his outside world. Quasi-mathematical operations to solve time/distance trajectory equations are suggested. Enactive, rather than verbalised, storage of information is discussed. Some practical consequences for training and equipment design are drawn in conclusion.

Introduction

Although train driving has been practised for 150 years, it is remarkable how little it is understood as a skill. Most people who give any thought to the driver as an operator believed his main task to be watching for signals. Signal observation is, however, only one of many tasks he actually performs and, in fact, his reaction to signals can be said to require relatively little skill. It is vital that the real skill of train driving should be understood so that the system designer can keep pace with advances in technology and help to promote safety even further than at present.

Until recently the formal training of drivers consisted of learning the Rule Book and of courses in appreciation of engineering and technical aspects of the traction machinery, i.e. locomotives, etc. After 6 weeks of this, the trainee embarked on some years of *informal* learning, during which he acquired, among

---

[1] (This paper is a modified version of an original report (Branton 1978) in which a task analysis and theoretical issues are discussed more fully. In this paper the section on the mental models of brake trajectories is new.)

other things, knowledge of routes. Officially, this was largely confined to memorising the line-side signals and other equipment specifically provided to inform the driver. The actual skill of trainhandling was absorbed during this period by a process of *incidental learning* akin to osmosis!

Methods of investigation

Task variables

The first step was to identify as many of the variables in the task as possible. This had to be done empirically over a period of years by trained observers riding on the footplate. Task variables were then listed and grouped according to their predictability in space and time; with predictability ranging from very low (fog patches, clouds, sunshine) to almost complete certainty (fixed line-side boards). Against this changing set of background variables, the driver controls the dynamic element in a large complete transport system: including lines of track, signal boxes and their operators, other trains in front, behind and across his path, etc.

Interviews and protocols

Interviews with over 200 drivers and inspectors, extending over some years, produced protocols and much indirect, anecdotal information. At the same time, it was found by observation that the work demanded *very accurate execution*. After many attempts to reconcile this accuracy with the vagueness implied by the anecdotes, it was concluded that the 'mental' component of train driving skill is almost wholly inaccessible to verbalisation by the unaided operator. Additional methods were needed to expose this component to investigation.

Behavioural observation

Prolonged and systematic observation showed that the driver, even with a second man in the cab, appeared to be relatively isolated. His control is one-dimensional, i.e. he can influence only speed and acceleration of the train. The task could be compared to process control, but such an analogy fits only partially, because most modern process plant is 'purpose-built', whereas the extent to which this is true of railways is debatable.

The driver's few overt control actions, accurate as they are, necessarily depend on task-related cues, but they usually have to be anticipatory because he very often can not see his target at all. When he can, he would often not be able to stop in time, if that was the required response. Our observations therefore included a search for perceptual cues in various alternative sensory modes. For example; it was possible to measure, albeit only crudely, the time lag between a brake application and the moment when all the brakes started to 'bite', by timing the interval between brake-lever movement and the forward inclination of the driver's trunk away from the seat-back. This lag, sometimes a matter of twenty seconds or more, depends on speed, type of train, etc.

The driver's behaviour is obviously goal-directed, but this exact goal at a given moment was the first object of investigation. For our study to be of any practical use, we had to try to catch at least a glimpse of what was being stored in memory and in what form; i.e. what inferences the man made at any given time about the states of the rest of the system; only on the basis of such information might it be possible to design suitable aids and direct training to enhance the driver's performance. Thus the first methodologically important decision was to *seek a purposive or prospective type of explanation* (Taylor, 1964) of drivers' behaviour, rather than regressing from an observed effect to 'a cause'.

The second point to make about the method derives from the previously observed difficulties experienced by drivers in expressing verbally what they thought they did. Bruner (1964) distinguished three internal information processing systems; 'enactive', 'iconic' and 'symbolic'. The problem of the form of skilled mental activity, taken together with the fact that drivers seem 'lost for words', leads to the proposition that some elements of their skill may be represented in memory 'enactively'; e.g. in the form of programmes of commands to skeletal muscles, rather than as verbal symbols. We should thus not be surprised at the men's inability to describe their work, but pay close regard to their actions as indices of their intentions and purposes.

The assumption of purposivity, or goal-directedness of skill, combined with critical analysis of the operator's stored knowledge, leads to what may be called an 'epistemic' approach. We watch successfully completed, skilled actions, then dissect our observations and ask at each stage what operations are needed to explain success. 'What would the driver *need to know* to succeed? What *did* he know from the outset? What *can* he know? What could he *not know* (without scanning the environment)?' And then we can ask him more about what he thought he knew. By this means we widen the scope of study to include behavioural observations and need not be ashamed to speculate and interpret behaviour purposively. If in the process this leads to potentially quantifiable information, so much the better.

Braking trajectories

Theoretical considerations about the space/time relationships of events in and outside the driver's head have been dealt with elsewhere (Branton, 1978), but some justification for going beyond the usual 'experimental' method is necessary here. The need to apply additional methods became clear when we first saw graphs of 'brake trajectories', prepared by railway engineers to illustrate changing train speed in a typical brake application on a locomotive. The

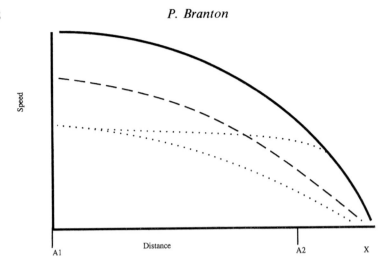

Figure 13.1 Engineer's braking curve (notional)

engineering problem is to ensure that the train stops at, or just before, a light
signal set at 'danger', or before touching the buffers at a terminal. In proportion
to the hundreds of tons of mass of the train and the speeds at which the
'projectile' travels on the open track, the point X, as we shall call the stopping
place, is a very precise location and the task of stopping a train just there is very
exacting. In practice, train drivers achieve very tolerable accuracy in this task, as
a rule. It is largely because of a continuous concern for safety as technology
advances, that machines and systems are planned which would supposedly stop
a fast train automatically at point X, or perhaps with the least possible human
intervention. The original reasons for our involvement in the matter related to
the human factors aspect of automation. However, as the study of the
information needs of the driver proceeded, it was realised that the problems
underlying the concept of braking trajectories were basic to a whole range of
aspects of many skilled human acts requiring explanation. Thus the immediate
purpose of the following remarks is to show that more help might be provided to
the driver in his task if attention were directed to braking point A, instead of
stopping point X, as at present. (See Figure 13.1.)

A typical braking trajectory

The braking curve usually drawn by the brake designer is based on his detailed
knowledge of the system and on his experience. It is a generalised, hypothetical
curve which says, 'The braking characteristics are such that the train will stop at
X, *if* the brake is applied at point A.' To the engineer, it appears as if point A
were fixed, just as a signal at point X is planted in the ground. And this would be
so in an 'ideal' world of certainty and invariability. But, looking with the driver's
eyes, the ergonomist finds quite a different situation. The driver operates in the

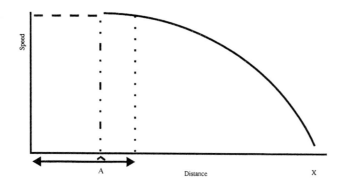

Figure 13.2 Driver's braking curve goal directed

real world and usually cannot actually see point X, his immediate partial goal. He applies the brake, in anticipation, well ahead of X, to reach it at zero speed. Not only does he usually not pass X, he also does not stop so far in front of the signal that it might change its aspect again by the time he finally reaches it.

Moreover, for the driver, point A has no fixed location on the ground. To him, A is variable because he has not only to consider the relatively constant brake characteristics at given speeds, but also many other variables; such as train composition and mass, the weather, etc., and even whether he is early or late. The driver must therefore hold in memory exact representations of these variables, as well as of all possible locations for the start of brake applications. The least one can say with certainty is that, as Figure 13.2 illustrates, point A will vary from one occasion to another over a range of locations. The driver makes his decision to apply the brake well before reaching the nearest point to him on the range of variability of point A.

This can be said to be a prime example of some of the psychological implications of goal-directed behaviour, awaiting study. The questions drawn to our attention by this schema of the driver's functions are: What must he *know*? What information can he *derive* from current perception of the environment? What is needed to explain his success in the task, carried out with a certain objectively measurable accuracy? The goal-directedness of his behaviour requires a different type of explanation from that typically provided by the engineers' analogy with a projectile.

A margin for safety

In the interests of safety, attention should be drawn to an asymmetrical oscillatory process occurring during the actual application of the brake, in addition to consideration of the point at which the application starts. Figure 13.3 is intended to show what a real braking characteristic must look like, if all the actually occurring imperfections of both the man and the equipment could be measured accurately. Ergonomists, aware of the potentially very great

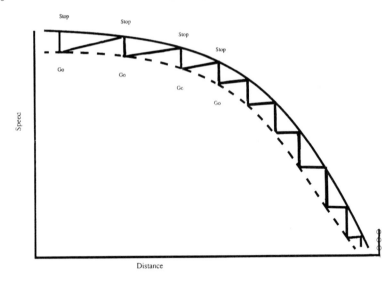

Figure 13.3 Braking curve (actual)

sensitivity and variability of the human operator, know that the apparently smooth curves illustrated in Figures 13.1 and 13.2 are in reality *'fuzzy bands'*. The actual performance of the driver is continually subject to perception and decision: brake on? brake off? The critical asymmetry arises from the fact that there must be no transgressing the solid line on the graph, or exceeding the real maximum permissible speed at any moment. The upper boundary of the zone is thus critical, the lower less so. It follows that the computations required, whether explicit or implicit, must not be simple averaging. Right down to the target X at the end of the braking curve, only a 'minus variance' can be tolerated.

Figure 13.3 is also meant to illustrate another phenomenon: the distinction between what we think of as perfect and what we actually experience. For it will be argued later that, to explain the almost infinite perfectability of skilled acts, internal representations of the world should probably be regarded as *idealised constructs*.

A model of the driver

The following model of the driver was assembled from recordings of running commentaries and interviews with drivers and inspectors, and from concurrent observations of behaviour.

Internal representations

The driver hardly ever sees a partial goal and certainly never the 'mission goal', or target, until it is too late to start implementing the appropriate response. He must therefore have representations of these goals in memory. To consider the nature of these representations, it is useful to get away from the usual explanatory concepts, such as long-term and short-term memory, which are often used as if they contained well formed and preserved sets of information items. It is preferable to speak of 'working storage' (Bainbridge, 1975), and thus make clear which information is retained by the man and not available directly by scanning the environment. In short, this is what he 'knows' from experience, plus what he acquires during a particular journey.

Not only does our purposive driver know his goal *from the outset*; he also knows from prolonged experience when and where along his path he has to seek further information on variables not as yet given. This knowledge determines or facilitates his scanning of specific parts of the environment for signals, etc. Among other things, he knows of course where and when he must act.

Besides the goals and sub-goals, he has internal representations of other points along the path and the evidence is that he connects them flexibly with the goals, using the former for orientation and anticipation. If this is the so-called 'route knowledge', it is not comprised merely of isolated snapshots of the scenery, but must consist of an ordered set of perceptual cues for active scanning and for motor action. The order is goal-related, probably in a back-to-front, count-down manner. (This, if true, has practical implications for information display.)

To speak thus of the nature of internal representations of goals or targets is no idle speculation. Only if the representations exist, and can be operated upon, is timely error correction possible; given the absence of direct perceptual feedback. The train drivers' very high success rate, in terms of punctuality, is rarely equalled in other transport modes. This can be taken as evidence for the likelihood that these goals and partial goals are very specifically and exactly represented in memory, in terms of time and distance. Again, it is unlikely that such representations are in symbolic, or even pictorial form; say, of the view at Euston three yards from the buffers. More appropriately, to cope with the time lag mentioned above, they would be of visual and other sensory cues a few hundred yards before these buffers. (This view of the task has fairly obvious implications for what has actually to be learned.)

Implicit operations

Now it is suggested that the operations performed upon the internal representations are *quasi-mathematical* and probably of two kinds: 'ideal' calculations and 'reality' corrections.

The evidence of exact execution warrants the inference that the driver intuitively solves time/distance equations for the whole journey, as well as for parts of his 'mission'. If he merely carried a 'pretty picture', of the track in his head, he would be neither as exact as he is, nor as flexible. Whatever else the operations and solutions of the equations may be, they are *idealised constructs*, or constructs of better-than-best-experienced-performance. *It is this urge to go beyond experience which comprises the train driver's skill.*

If expressed verbally, the simplest form of the driver's subjective forecast might be as follows: 'If all had gone well, I should be seeing this scene at three minutes to six; then I need do nothing. Anything else is an error and I must either speed up or slow down'. This means that the stored representations are the *consequences following from his anticipated perceptions for his subsequent actions.* This way of putting the case is perhaps not easy to grasp at first, but it allows parsimonious explanations of the storage system, as well as of the open-endedness of purposive behaviour. A stored action can be added to the skilled man's repertoire without his necessarily having carried it out. (Pribram, 1976).

From the subjective forecast hypothesis that skilled storage consists, not just of anticipated external events, but of probable actions consequent upon those events, it follows that:

(a) there can be no one-to-one relationship between percept and action. Actions are not automatically cue-triggered, and simple reaction times to perceived line-side signals are thus not appropriate as measures of skilled performance.

(b) there must be a choice among stored strategies in a repertoire, probably made very centrally. The decision clearly appears to be based on the significance which the changes perceived outside may have for the goal of the implicit calculations.

The man could not solve, as he does, the postulated partial time/distance equations unless he took account of the state of variables as he encounters them *en route*; i.e. by empirical, perceptual, information gathering. On perceiving changes in any of the variables which *he thinks are significant*, he re-calculates his subjective trajectory, only to test the results against further perceptions.

Probably the solutions to the postulated equations are expressed in sensations of speed, acceleration and deceleration. How does the driver 'know' these parameters? With the equipment known to be available for internal operations, this knowledge would require complex processing of vestibular, kinaesthetic, acoustic, and even peripheral visual information. This is of course a much richer source of information and has more 'redundancy' than is provided by the present machine displays. (Indeed, it is known that the instruments in the cab are of mainly confirmatory value.) Each element of these multi-sensory data sets contributes a well defined share to a *band of tolerances* from which, normally, little obtrudes into the driver's consciousness.

The edges of the tolerance band are at the upper and lower limit of expected performance. The upper limit is what the driver considers compatible with his conscious feelings of safety and comfort for his passengers and for the integrity of the whole railway system. He is equally well aware of the lower limits of expected performance, below which the successful achievement of his goals is dubious. For instance, being late at one stage of the journey will make him exert considerable effort towards approximating the upper limit of his available resources during the next stage.

Skill assessment and experience

Throughout the interviews and conversations with the highly practised operators, the outstanding impression gained was of tremendous confidence in their ability to judge the skill of others and to recognise potential performance. Incapable of verbalising how they arrived at their judgement, they nevertheless claimed intuitive understanding. If this were a form of pattern recognition, that recognition is certainly not of graphic patterns, but rather of timing and smoothness of movement. It seems to be anchored in the special character of 'enactment'; in other words, someone who has successfully carried out a task is best able to recognise the *degree* of skill exhibited by other practitioners. Evidence for the existence of enactive programmes can readily be observed in the empathetic behaviour of members of an audience at, say, boxing matches, or football games.

The nature of this skill

Having described some of the activities we observed in train driving, and some of the internal operations we inferred therefrom, we can draw together some points about the essential nature of this skill. Among the many features relevant to ergonomic analysis, we may select somewhat arbitrarily the following qualitative and quantitative aspects. To carry these descriptions beyond a mere listing of the relevant features, we will point to differentiation in degree of skill, where possible.

Qualitative aspects

Anticipation: The operator could hardly succeed without the ability to anticipate the values of a large number of variables. He must, furthermore, be aware, however vaguely, of the consequences which each change in variables will have for his prospective decisions. Above all, he must judge the relative significance of variables. The greater his anticipation and the more elaborate the consequences, the higher the skill.

Internal representations: The operator's internal model of the external system in which he works is probably a perfectionist's idealised construct; not quite like anything he has ever actually experienced, but rather a 'conditional approximation'. Otherwise he would not possess a notion of his goals which is sufficiently specific to permit the detection and correction of error-deviations. If this is so, skill would rest upon alertness to change, or sensitivity to deviations from his goal-notion.

Reality testing: A continuous interchange takes place between the operator's inner representations and the outer, real, world. This interchange consists of motor activity and the focusing of attention to specific sensory receptor systems, e.g. visual scanning, and the setting of joint and muscle receptors to confirm the arrival of expected events. The objective of this interplay is to achieve relative congruence between internal model and real world states and the skill shows in the confidence the operator exhibits in his forecasts.

Quanitative aspects

In principle, it should be possible to take these aspects of skill into the laboratory, simulate them appropriately, and experimentally develop them as parameters of the skill.

Forward extension: This would probe how far ahead, both in time and space, the driver extends his internal operations at any given moment.

Size of repertoire: This would test, say, the number of sub-routines or alternative actions available to the driver to achieve his goals and also measure any eventual increases in number occurring with experience. The strategies to which he habitually resorts could then be described as his 'style'.

Motor sensitivity: This would test the driver's capacity to discriminate deviations from his own norm of performance in, say, brake application, and the increasingly fine grading of error correction.

Reality adjustment: This may be measurable by comparison of the driver's subjective probability estimates of events with actual, real-world occurrences.

Practical recommendations

The consequences for the enhancement of skill and refinement of training, outlined in the foregoing views, can now be stated briefly. Only a few of the possible, purely practical and economically viable measures, will be considered here. One could, of course, speculate on an enormous range of technically sophisticated (and expensive) ways of replacing men by machines. But at present even the most progressive technologists would replace only *some* men by *some* machines, and so the problem of interface design for safe and skilled performance remains.

Aids to orientation

From the finding that the driver treats route knowledge as an integrated whole, composed of many more variables than are, or can be, provided by the railway engineers, it follows that he may not necessarily need *more* information. Rather it may be that information needs to be *better presented*, i.e. in forms increasingly appropriate for the purpose of orientation and finding his position along the track unambiguously and in forms which are easily recognisable in all kinds of weather and visibility conditions. The recognition and discrimination of invariant landmarks is an obvious factor to be learnt and practised. At the same time, provision of uniform equipment, such as posts or gantries for the overhead cables, even if they are conspicuously numbered, will not necessarily enhance orientation, since it requires further mental effort by the driver in order to connect a number with its geographical location. To aim at increasing the probability of correct orientation is to express only a principle, the application of which remains to be tested empirically.

Enhancement of anticipation

Now that we have a hypothesis concerning the nature of the driver's anticipation, we are in a better position to attempt the design of driver aids. The model of *'subjective forecast of consequent action'* suggests that storage may be in the form of a repertoire of programmes of one's own actions. It follows that:

(a) Information displays might usefully be designed to fit this form of representation. A predictive format (Kelley, 1968), and account taken of habits and movement stereotypes, are likely to reduce time-stress in decision-making.
(b) Because the stored programmes of the less probable contingencies are more likely to decay through lack of practice, safety requires a deliberate shift of attention to them. Organisational provision of periodic refresher courses, designed to exercise such normally infrequent decision-making, might strengthen scanning capability and increase perceptual sensitivity.

Improvement of motivation

A lasting impression remains, from the field study, of the relative isolation of the driver in the cab. The effect may be ascribed to two factors. On the one hand there are the challenging and motivating factors of independence and the responsibility of being in sole charge: on the other, the limited feedback on success or error in performance. On long journeys the achievement of, at least, partly self-set targets can be a source of satisfaction, through self-confirmation. If two-way channels of communication, e.g. by radio, could be established and made more informative than the present system, the relief from social isolation would, perhaps, not only improve motivation, but also permit better anticipation and understanding of the state of the rest of the system.

Enriched environment

The ergonomics and environmental requirements for cab design have been discussed elsewhere (Grant, 1971, Branton, 1970) and only a few brief points will be made here. Climatic conditions at the work place – heating, cooling, humidity and air change – will obviously affect skilled operation, as they are likely to make for greater or lesser attention and vigilance. It is, however, not always appreciated that, while drivers are now certainly better protected from the weather than they were in the days of steam, their sensory environment has been impoverished to some extent (see Endo and Kogi, 1975). Some sensory deprivation may arise from two sources: lack of perceptual cues, and uniformity of climate in the cab. The former are needed to provide visual, acoustic and kinaesthetic information used in orientation and assessment of speed, etc. In the latter case, an excessively uniform heating and ventilation system tends to produce attention-reducing monotony and may lead to all manner of attempts to vary the environment provided, including opening windows or blocking ducts; thus defeating the design purpose.

In any case, the most useful enrichment that can be given to the driver is equipment which is well-designed and well-maintained, and validated in his terms; he can then demonstrate his competence and have the confidence that he is in effective control.

References

Bainbridge, L., 1975, The representation of working storage and its use in the organisation of behaviour, in W. T. Singleton and P. Spurgeon (Eds) *Measurement of Human Resources*, pp. 165–183, London: Taylor & Francis Ltd.

Branton, P., 1970, Train drivers' attentional states and the design of driving cabs, *Invited paper to 13th Congress (Madrid), Union Internationale des Services Medicaux des Chemins de Fer*, Brussels: UIMC.

Branton, P., 1978, The Train Driver, in W. T. Singleton (Ed), *The Study of Real Skills*, Ch. 8, Lancaster: MTP Press.

Bruner, J. S., 1964, The Course of Cognitive Growth, *American Psychologist*, **19**, 1–15.

Endo, T. and Kogi, K., 1975, Monotony effects of the work of motormen during high-speed train operation, *Journal of Human Ergology*, **4**, 129–140.

Grant, J. S., 1971, Concepts of Fatigue and Vigilance in Relation to Railway Operation, in I. Hashimoto, K.Kogi, and E. Grandjean (Eds) *Methodology in Human Fatigue Assessment*, London: Taylor & Francis Ltd.

Kelley, C. R., 1968, *Manual and Automatic Control*, New York: J. Wiley.

Pribram, K. H., 1976, Self-consciousness and intentionality, Ch. 2 in G. E. Schwartz and D. Shapiro (Eds), *Consciousness and Self-regulation: Advances in Research*, London: J. Wiley.

Taylor, C., 1964, *The Explanation of Behaviour*, London: Routledge & Kegan Paul.

Chapter 14

The use of critique in meta-psychology[1]

Paul Branton

Society for the Furtherance of Critical Philosophy, Ratio, 1982, **24,** *1–11*

Introduction

I propose to pay my respects to Leonard Nelson, whose 100th birthday falls this year, by presenting my own understanding of the relevance of his work to the philosophy of mind. The critical method, to which Nelson was committed, attracts me, as a psychologist, because of its applications to research methodology as well as to the formation of theories about functions of the human mind and body. I see great benefits for both psychology and philosophy if they could pool their present knowledge to concentrate on the problem of which human faculties and capabilities are really the most basic ones. Even today, many theorists go back to Aristotle, Aquinas and Hume and adopt their views seemingly without taking account of how much more complex processes like perception and cognition are than these savants could possibly have known. As a result anyone working in the fields of social science, in education, psychiatry or any of the other applied human sciences must be utterly confused by contradictory theories and thus in urgent need of up-to-date conceptual frameworks.

Nelson, extending the Kantian tradition, pursued the argument that possession of knowledge of reality, necessity, causality and other notions connecting events, must come before experience. Indeed, a person must first have the capacity to turn single events into coherent experience. Nelson's special contribution concerned the linking of epistemology and psychology without either dominating the other: every metaphysical proposition must be matched by another statement, namely that this is psychologically possible. Conversely, psychological phenomena can only derive meaning from metaphysical interpretation. For Nelson, this implied the possession of a mental capacity for immediate knowledge. He saw that this does not mean a person need be immediately conscious of possessing such knowledge; but the psychological aspects of this insight were not as fully developed as I would wish.

The first theme I want to discuss in this article is about the implications which immediacy of knowledge has for true autonomy of behaviour, a central question of all psychology.

Another central concern of psychology has always been the gap between knowing and doing. That gap is usually filled by some mystical kind of mechanistic concept like instinct, which forces human behaviour into a strongly deterministic framework, thereby denying the possibility of choice which surely is a necessary condition of autonomy. The evidence now points away from any rigid determinism and, in the attempt to shape theories accordingly, a movement towards a more specifically human 'cognitive' psychology has come into being. My second theme here will try to carry this renewal further by briefly outlining Nelson's Theory of Interests, which is explicitly psychological and is a growing point for my own work. It seems to me that the gap between thought and action might be bridged by such a theory, provided certain of Nelson's ideas about basic human faculties are reconsidered in the light of modern knowledge on emotions and brain functions. In the process of carrying further our understanding of the above two themes, the value of the critical method will become apparent. As Nelson was devoted to a revival of the Socratic Method in practice, it would be a fitting tribute to show how this method could be extended to the teaching and practice of psychology.

Topicality

For a psychologist, trying to combine practical experience with theory, Nelson's work is now more topical than when it was written. While many of the psychological theories of the recent past fail for lack of philosophical grounding, his efforts on method, in particular, bear directly on the foundations of a consistent anthropological psychology. For Nelson this necessarily included the concept of autonomous man living in a complex society. Since his day, factual scientific knowledge on functions of the human brain and mind, on mental activities and on the nerve cells, has increased vastly, yet the same basic problem of the sheer possibility of inner experience and of having knowledge still demands understanding and explanation. Nelson not only perceived and upheld the value of J. F. Fries' contributions to philosophy, mathematics, medicine and psychology, but strove to bring philosophy and modern scientific method closer together. He hoped to put the study of man's inner life and of ethics on a more scientific basis. The earnest of this intention is expressed by the dedication of his major work to his teacher, David Hilbert, the father of axiomatics. Nelson was definitely no believer in absolutes. In one of his first major works, a review of epistemology, he concluded that a theory of knowledge was inevitably bound to fail; its propositions had to be hypothetical and were therefore not on a level of abstraction high enough to avoid circulatory and infinite regress. A first statement needs to be in the apodictic mode. The only method by which to grasp non-mathematical, non-demonstrable 'truths' of inner experience and to arrive securely at such first statements was critique.

Nelson would be glad to find philosophers now no longer preoccupied with logic and language *per se* and turning again to practical puzzles of cognising the world. Equally he would welcome the sight of psychologists engaged again in meta-psychological speculations, thanks to cognitive psychology, and no longer regarding 'mental' as a taboo word. Such speculations now combine psychology systematically with neurology and chemistry. Re-thinking is under way, as will be apparent from the following quotation from Dennett, representative of the English-speaking main stream of current thought on the mind-body complex of problems.

> One can ask how any neural network can possibly accomplish human color discriminations, or one can ask how any finite organic system can possibly subserve the acquisition of a natural language, or one can ask, with Kant, how anything at all could possibly experience or know anything. Pure epistemology thus viewed is simply the limiting case of the psychologists' quest, and any constraints the philosopher finds in that most general and abstract investigation bind all psychological theories as inexorably as constraints encountered in more parochial and fact-enriched environments. (1979, 162)

Seventy years earlier Nelson too had argued for an intimate bond between epistemology and psychology:

> . . . for every metaphysical statement there must be an equivalent psychological statement, namely a statement about the grounds of possibility of the knowledge expressed in the metaphysical statement. Regarding its validity, this psychological statement stands in a relationship of reciprocity to its metaphysical equivalent such that both statements can only claim to be founded on truth *together* or must be rejected *together* as false (1973, II, 367) (Transl. PB.)

My own line of thought would go on like this: We begin by accepting Kant's transcendental argument that humans cannot know the 'real truth' – whatever it may be – with *absolute* certainty. For the possibility of 'knowing anything at all', we must therefore ultimately resort to sources of knowledge which lie *within* ourselves. Such knowledge as we have, we possess only as actually thinking beings who process by their own effort whatever information they receive from outside. From this we may validly conclude no more than that the most basic human faculty is the ability to think.

The Kantian 'I think' can become the major premise for what follows on the immediacy of knowledge and human autonomy. However, before we can develop these concepts, we must deal with apparently conflicting assertions.

Basic mental faculties

When asked which human mental faculty is most basic, modern theorists almost invariably first turn to Aristotle's trinity: cognition, conation and willing. I maintain, in the light of recent brain research, that this conventional division is wholly unreal and unnecessary. To think of the three as separate things or mental states, perhaps even having their 'seat' in specific and separate locations in the head is not only inappropriate to many practical applications in clinical, educational, experimental and social contexts, but also bedevils psychiatry and brain surgery when irreversible intervention might be considered. Whatever the most basic *mental activities*, the Aristotelian three – knowing, feeling and deciding – look more and more as one and the same ongoing process, a dynamic 'steady state' in parallel with continuous creation. We should not separate them into discrete events except for convenience of description.

It is interesting that Socrates, without the benefit of modern brain research, came to the same conclusion. When he doubted that a man will *knowingly* commit a wrong, he most probably meant by knowing more than mere perception and belief in the existence of a thing. That often quoted remark becomes less puzzling when 'knowing *what is right*' includes 'knowing *what to do*' as well as actually *doing it* and these are not thought of as separate.

In the normal scientific way concepts would be defined at the outset of discussion. But psychology cannot be axiomatised, as can some branches of mathematics. It lacks the canonical laws of some physical sciences. It makes no cast-iron inferences from an observed effect to its 'only possible' cause. Even an exact operational description by itself of how the brain supposedly works is not likely to satisfy us when the question is what is most basic to the human mind. To accept Aristotle's or Aquinas' or Hume's view uncritically would be to start in the dogmatic, not the Socratic way. The Socratic is the discovery method in which definitions are not the starting point, but the end product of discussion, however tentative and provisional the outcome may be. This is the approach of critique. To ensure relevance to real-life problems, it must start from views that are actually and sincerely held so that there is at least one genuine protagonist in the discourse. A brief critique of cognition usually relates perception to conscious knowing, thinking and understanding, to intelligence and the intellect, to linguistic expression, to the use of concepts and to 'handling' symbols for problem solving and the like. Philosophers too use the term in their theories of cognition. But, as Bruner (1980) shows and many other psychologists frequently remind us, we know more than we experience and we experience more than we perceive. There must be more mental activity besides *conscious* cognition and we must have more knowledge beyond that of which we are conscious at a given time. Supposing some unconscious knowledge is immediate knowledge, might it not be that such knowledge is originally non-cognitive? Might it not be non-conceptual and non-linguistic?

Thoughts without concepts. The existence of pre-conceptual thought has long been postulated in various guises[2] and, in my view, there must be a stage in thinking about something before its proper concept is formed. Indeed I would go further and postulate explicitly the existence of a kind of thinking which need never rise to the clarity and consciousness of a specific concept. If a concept of something can have a symbolic representation, in terms of natural language or other intermediary transformation, e.g. into logic or 'artificial intelligence' computer code, it is *ipso facto* consciousness-bound and expressible. An unconscious concept seems to me a contradiction in terms. Therefore, if some thinking is non-conceptual, I doubt whether it can be called 'cognitive' mental activity without causing confusion. It may be difficult to speak about this other kind of thinking, but the possibility that it exists cannot be ignored. There is in fact evidence for a kind of thinking-by-doing and for 'thinking with the hands', which I presented elsewhere (Branton, 1978, 1979).

That something of which one is not conscious may be raised into clear awareness may seem an epistemic paradox. A psychiatrist might try to help, by offering from the outside a guess at what might be there. But if I do not know what is in my own mind, how can anyone else really know for sure? In the end, the sustained raising of any unconscious knowledge must be an *inner* process. The person himself might do it by exercising critique on his own non-conceptual knowledge when he has an intimation of some item of knowledge-before-conscious-experience and then inquires into his pre-assumptions that made this possession possible. The process of bringing immediate knowledge into awareness therefore presupposes the autonomous person as a thinking being.

In my view, to be thinking is to be forming implicit or explicit propositions to put to oneself. This, I believe, is the next most basic mental faculty. It necessarily requires the presence of a Self which is, at least sometimes, conscious of itself. Kant and Fries proposed that there can be only one 'pure' proposition and that it had to be *reflexive*, i.e. experienced by oneself. Kant expressed it as 'I think'; Fries called it 'the thinking Ego in pure-selfconsciousness' (1798, 1968, 219). Since both these are open to a narrow interpretation of thinking as a cognitive, even a verbal, activity, I prefer to say just that 'I have a mind'. More is not needed at this stage.

Without the thought of "It is MY mind which is now thinking", none of the subsequent process could operate fully. This then is the "substrate of inner experience" (Fries) onto which all propositions and particularly the predicates of outer experience are mapped. These propositions to the Self are implicit and need not be in natural language form; neither need they be in symbolic or pictorial or even in iconic form. Self-knowledge can take many forms; even my thumb performs the role of my Self when the index finger presses the pen against it. This knowledge is so immediate that I can think of it only because I know some neuro-anatomy; but nobody stands between me and my thumb. If they could be verbalized, such propositions might be, 'A change has occurred out there,' or 'This lamp is round and red,' or 'Life is great', or 'The pen is slipping'. They have subjects, predicates, objects and verbs linking the elementary parts of propositions together in thought.

Before a bridge can be built from knowing to doing, we must go into the role the propositions play and how they allow the person to take in information from the world around him, especially the social world. If the only pure proposition states *self*-consciousness, that person might be autonomous but would also be egocentric. Other implicit statements must therefore let the outer world in. Their role is to *relate* or *separate* events or features and attributes of objects of the world *in our thoughts*. I must therefore postulate two further basic human mental faculties, complementary to each other: to effect relations or conjunctions and abstractions.

> Among all notions (*Vorstellungen*), the one of conjunction is the only one not given by objects, but can only be carried out by the subject himself, because it is the work of his self-activity. (Kant, *Critique of Pure Reason* (trans P.B.))

I suggest that the implicit propositions put to the Self have the same categorical characteristics as explicit ones. Equally, they propose to the Self an inherence of *lawful* character for events recurring regularly. They thus *relate* events as effects to causes, or existence to constancy of matter.

Proposed relations and abstractions: The psychological process of relating is perhaps more easily understood than that of abstraction. Hume proposed laws of association by repeated proximity of two external events to explain the sequential nature of mental activity. Not only would the conjunction of events be recalled, but we would also come to expect the conjunction to have the same effect each time. The assertion of causal lawfulness based on experience alone provoked Kant to expound the need for postulating the existence of *a priori* synthetic propositions. Similarly Nelson insisted that frequent coincidence or regularity was not enough to ground our ideas of necessity and lawfulness, which must come from an exercise of critique upon our presentiments and presuppositions. I suggest this is at the core of the problem of 'pattern recognition' which now vexes Artificial Intelligence scientists who find that computers cannot equal humans in this respect.

The difficulty of describing the psychological process of abstraction has been much reduced lately by advances in psycho-physiology, when it was shown experimentally that during a very early stage in perception an object is analysed and each of its features or attributes 'abstracted' for internal representation in a separate cell or group of cells. For example, the specific wave frequency band of the colour 'attributed' to an object activates one or a group of cells, the straightness of outline of its shape another group, the obliqueness of that line yet another, and so on. It is therefore possible to say that, *in our heads*, the sum of the predicates of all the propositions we implicitly make upon perceiving an object IS this object *for us*.

> The predicate of a proposition is always a form and a particular sensible thing is nothing else [in our heads] but the common meeting place of a number of predicates. . .Apart from these predicates, the thing. . .has no independent reality, and if we know all the forms in which a thing participates, there is nothing more to know about it. (Burnett, 1914)

It may surprise that this quotation is actually Burnett's description of Plato's Forms as a philosophical doctrine, besides fitting the physiological facts. Could the facts and the theory have that much in common by accident? Even if the

Theory of Forms is regarded as idealistic, this formulation of it and the factual arrangement allow for an infinitely perfectable match between the inner and the outer world rather than linking the two rigidly.

The finding that predicates are abstracted from their objects has profound consequences for understanding the mind. It helps to explain how a vast collection of things can be internalised, stored with economy and reassembled again in recall. Things may thus share attributes with other things and their representation need not be concentrated in one point. This could account for the phenomenal flexibility and resilience to damage of mental 'images'. If abstracting and relating are paired, dynamic mental operations, representations of things are continuously construed and re-construed, and related by persistent checking back to the perceived 'real world', as it were by zooming in and out of focus. In the same way, critique modifies abstractions by constant reference to 'reality'.

Nelson pointed to yet another important aspect of the psychological process of abstracting: its social function. As mentioned earlier, it would be possible for an autonomous person to be purely egocentric. If I were to relate everything I perceived wholly to myself, other beings could forever be my objects, without a personality of their own and mere means to my ends. (The behaviour of psychopaths is sometimes described thus.) Even the most primitive social interactions could not take place without an ability to abstract from the self-conscious Ego. However, as soon as that 'thing-out-there' is registered as 'another person', we can identify ourselves and 'come to see ourselves as one among others'.[3]

This brief outline of a critique of 'cognition' can only point into the direction which a re-conceptualisation may take and indicate some of its implications.

Emotional and reasonable acts

Having considered the possibility of immediate knowledge, my second theme concerns the theoretical possibility of 'non-immediate' action, that is action which is not so impulsive that reasoning had no chance to intervene. The process must be such that the person is neither wholly stimulus-bound nor bound to respond mechanically once in possession of knowledge. At this point, where *choice between courses of action* must enter, most theories fail to give a satisfactory explanation of how reasoning is compatible with the forceful decisiveness required for any act. Here too, critique can help when it sets out from statements about facts and seeks the principles which must be assumed to operate so as to explain these facts.

It is one fact that we sometimes deliberately choose one particular course of action, both in regard to means and ends, and we cannot therefore be regarded as pure automata. However, knowledge alone does not always issue in action. When we do act, there must also be some prime force within which moves us. Where do we find this force? In his critique, Nelson points to another fact, namely that we have emotions of pleasure or displeasure upon perceiving an object. These emotions are immediate 'feelings of worth' as well as *desires*. Their varying strength constitutes that basic motive force. A desire is an expression of one's *interest*. To interest oneself in something is to *confer upon it a value* (or lack of value). I would regard these emotions as *crypto*-valuations because one need not be immediately aware of a particular value judgement or interest. As emotions, they share the quality of inner experience with Aristotle's basic faculty of conation, and with other 'activity' terms like feeling, striving, and orexis. As valuations, their immediacy makes them personal sources of action which might at later stages become conscious desires, motives, interests and values by a process of *adumbration* and reflection (cf. Weimer, 1977). No doubt, these are at first subjective valuations, at least some of them may even be sheer 'prejudices', yet they are capable of being modified into 'trans-subjective', not to say objective, values.

(There is now strong psychological evidence for saying that we immediately invest *every* bit of information we receive with an emotional charge. This charge is not the sense impression of Hume's account of the process; one attaches it to the sensory input by tacitly asking onself, 'What is this *to me*?' (cf. Pribram, 1971)).

Interestingly enough, any arbitrary line between basic mental faculties becomes blurred in the course of adumbration and the conative can become cognitive. One may recall Socrates' concept of the Self, in which apprehending the right and desiring it were yoked together by emotion, just as knowing the wrong meant abhorring it. Only when *nous* and *orexis* diverged would there be trouble.

I have no doubt that the combination of the evaluative and the motivating functions of emotions gives this theory considerable explanatory power. Although he does not make an issue of this particular point, Nelson succeeds more nearly than other meta-theorists in consistently grounding a *purposive theory of human action*.

The foregoing thoughts are largely the results of my attempts to apply the critical method to psychological research. If I were to try and formalize the description, I would begin by saying that it is meant to augment or sharpen, rather than replace the experimental methods at present practised. The pre-assumption of purposiveness of behaviour has a decisive influence on four stages of a research project. The first is the requirement of an initially detached and naturalistic observation of the behaviour under review. This is to ensure as far as possible a realistic base. It requires all the skills and ingenuity of the researcher to be unobtrusive and non-invasive with his subjects, so as not to foist his preconceived hypotheses upon them and so obtain merely self-confirmatory results. Having obtained protocols and, preferably permanent, records, the second stage is to analyse them and to conjecture that the person somehow 'knows' his own purposes. It is then for the researcher to find out on what implicit hypotheses the subject operates. The question is not, 'What made him do it?' as this implies outside agency. Instead, we ask, 'What is he trying to achieve?' or 'What are, or were, his *purposes* or reasons for acting as he does or did?', assuming that he is his own 'cause'. Then we ask ourselves the more difficult question, 'What information from outside and from *his own feedback* would be necessary and sufficient for him to operate and act as he does, or did?' The researcher should, at this stage be mindful of the originally obscure and subconscious nature of the person's knowledge and information processing. Depending on the topic under consideration, the Subjects' own introspective reports must be taken into account, though they can only rarely be taken at full weight. They are of particular interest when they emerge from the appearance of whatever objective measurements are taken concurrently with observations.

The third stage is the one at which the partial hypotheses formed by the researcher are taken to the laboratory or back into the field and salient aspects subjected to controlled experimentation for confirmation or refutation. The fourth stage consists of re-integration of the part-hypotheses into a whole and the formulation of a generalized solution.

In conclusion, I hope that the critical method recommends itself to psychologists, as it does to philosophers. I also hope that what is now termed 'cognitive psychology' will develop into a wider study, dealing with what 'exists' in the mind. That will be more than what is conscious, intellectual, conceptual, linguistic, and symbolically representable. To confine mind to these five would be a grave mistake. If I may speculate, I would think that the term 'cognitive' was revived with the intention of re-asserting a psychology in which mind was allowed an existence, albeit in a different and more limited form than the old 'pure' mentalism. Perhaps, we may now feel our way to a facultative and anthropological psychology, i.e. one in which *structure qua body* and *function qua purpose*, as distinct from function qua automatic mechanism, is given more consideration than causally deterministic explanation. Maybe the less interesting, backward-looking explanation will be replaced in due course by purposive determination, treating the person as a forward-looking, reasonable being.

Notes

1. Paper reprinted by permission of Blackwell Publishers.
2. I have a collection of expressions, or circumlocutions, to which I add every time I come to the crucial point in a philosopher's work. These are most often found tucked away in a footnote. Examples are: 'unstudied utterances' (Ryle), 'non-observational knowledge' (Anscombe), 'pre-reflective consciousness' (Sartre), 'proximate propensities' (Sellars), 'propensities of the self-conscious mind' (Popper and Eccles) and even 'feed-forward' and 'precognition' (seriously suggested by cyberneticists, e.g. L. Young). I myself would recommend 'intimation', 'presentiment' or 'premonition', terms explicitly used by Fries (1828) and Dennett (1979, 165).
3. I owe this way of putting it to Angus Ross and a paper which he gave to a seminar in Oxford in June 1981.

References

Branton, P., 1978, 'The Train Driver', in Singleton, W. T. (Ed.) *The Study of Real Skills I: Analysis of Practical Skills*, Ch. 8, Lancaster: MTP Press.

Branton, P., 1979, 'The Research Scientist' in Singleton, W. T. (Ed.) *The Study of Real Skills II: Compliance and Excellence*, Ch. 13, Lancaster: MTP Press.

Bruner, J. S., 1980, *Beyond the Information Given*, London: Allen & Unwin.

Burnett, J., 1914, *Greek Philosophy: Thales to Plato*, London: Macmillan.

Dennett, D. C., 1979, *Brainstorms: Philosophical Essays on Mind in Psychology*, Hassocks, Sussex: Harvester Press.

Fries, J. F., 1798, Ueber das Verhältniss der empirischen Psychologie zur Metaphysik, Vol. 111. Chs. VII–XI, p. 156–402. *Psychologisches Magazin*, Jena: Cröcker Verlag.

Fries, J. F., 1828, *Neue oder Anthropologische Kritik der Vernunft* (1828–1981) Aalen, W. Germany: Scientia Verlag.

Kant, I., 1968, *Kritik der Reinen Vernunft*, Vol. III. p. 107: Akademie Textausgabe.

Nelson, L., 1973, Ueber das sogenannte Erkenntnissproblem, Vol. II, *Gesammelte Schriften*, Hamburg: Meiner Verlag.

Pribram, K. H., 1971, *Languages of the Brain: Experimental Paradoxes and Principles in Neuropsychology*, Englewood Cliffs, NJ: Prentice-Hall.

Ryle, G., 1949, *The Concept of Mind*, London: Hutchinson.

Weimer, W. B., 1977, 'A conceptual framework for cognitive psychology: Motor Theories of the Mind', in R. Shaw and J. Bransford (eds) *Perceiving, Acting and Knowing*, London: Wiley.

Chapter 15

On being reasonable

Paul Branton

(1983) 'Vernünftig sein' in Vernunft, Ethik, Politik: Gustav Heckmann zum 85, Geburtstag (eds D. Horster and D. Krohn) 1983 Hannover, SOAK Verlag

At a time when the world seems to have gone crazy, the cry from some quarters is 'Let us be reasonable – not coldly, calculatingly rational – just reasonable'. Being reasonable is, above all, being thoughtful, in contrast to being impulsive. The thoughts are, however, not just idle ones but thoughts with a view to action, practical rather than merely contemplative. Such a demand would hardly be possible if one could not ever be reasonable. It must be assumed, then, that most people, if not all, CAN be reasonable at some time or other. We take ordinary reasoning so much for granted that there are few very detailed accounts of it and even fewer are satisfactory. Trying to find out how exactly it must work in practice, I came to find it a most remarkable activity, because, strictly speaking, reason allows us to manipulate our future before it actually happens.

The topic can be studied in two ways: by setting up psychological and physiological theories based on the fact that, whatever reason may be, it must dwell inside a head and is therefore a matter of 'inner life' which a fact-finding science can explore. Or it can be approached as a philosophical inquiry, since reason is something both abstract and generally expressible in language. I take the view that both disciplines can and do deal with one and the same set of events and that both can legitimately claim to be equally true – provided both are applied critically and neither contradicts or excludes the other. Then each will illuminate the objective in a landscape from a particular angle. No single discipline can hope to treat so complex an objective exhaustively, yet the focus remains the same – reasonable mind. One discipline views it as within an individual person, the other as a common factor *between* persons. (This notion of a 'split-up of truth' was – to my knowledge – first clearly expressed by Fries in his *System of Metaphysics* in 1824.)

Keeping to common usage as much as possible, I am speaking here about Reason, without the definite article, *Vernunft* in the Kantian sense, and about reasoning, as distinct from intellect or intelligence, *Verstand*, which I consider as included in and subordinate to the former. This is also different from 'having reasons *for* something'. Reasons in the plural are usually means of explanation

and thus instrumental. I shall treat reason and reasoning as the ends of thinking
and the beginnings of action.

From a philosophical viewpoint, an effort of imagination is urgently required
to overcome a prevailing proneness to think in black-or-white, in disjunctions as
if they were complete when, in fact, they may well be incomplete. There may be a
third alternative, not grey, but brightly coloured. I am therefore resolved to
adopt a habitually critical stance toward dichotomies. Furthermore, it would
not help my argument if one were to get immediately involved in labelling each
thought in terms of the currently fashionable conceptual disputes, e.g. mater-
ialism/idealism/realism, or monism/dualism/pluralism, or parallelism/inter-
actionism/transactionism, or subjectivism/objectivism, etc. Such prior mental
sets – some would call them prejudices – actually prevent new combinations of
old ideas and give arguments like the present little chance to show how fruitful
they can be.

A further problem in philosophical and psychological discussion is presented
by the language difficulty to pin down events and dynamic *processes* in space
and time in verbal description of thinking and reasoning. Logically exact
language may be suitable for dealing with static things and passive properties
possessed once and for all; processes, especially in 'real life' are better grasped by
fuzzy logic and infinitesimal calculus.

Scientific verbal statements about reason are therefore not tenable with the
same degree of certainty as those about palpable matter. Reason, as a process, is
clearly abstract, done by thinking and only by this means. As an INNER
process, its existence must be a matter of conjecture for anyone other than the
reasoner. Knowledge of another's internal activity cannot be as certain as strict
scientists – especially empirical ones – would wish. Is it therefore 'less true' ? Of
course not, it is merely 'less absolute'.

Detaching the observer

In this attempt to bring together philosophy of mind, psychology and
physiology in a systematic approach to the topic, evidence from 'objective'
observation, experimentation and measurement will be used, even if this
provides only indirect and limited support. More importantly still, given that
reasoning is an *inner* process, the observer of people must not ignore the special
relationship he willy nilly adopts to his subject. The man in the street cannot but
say to himself. 'He looks like me, therefore he must be like me in one or other
respect and therefore he thinks as I would in the circumstances'. This has been
called the 'irresistible analogy'. That person may use the analogy quite happily
throughout his life without ever worrying that his judgements are founded on a
mere analogy from appearance to inner quality. The critical scientist, who
happens to speculate about that man must, however, be aware of the weakness

of analogies. Indeed, the sophisticated observer who hopes to rely wholly and solely upon the evidence of his own senses and who, at the same time, abhors introspection as unscientific, must be argued with *ad hominem*:

> You are mistaken if you believe you can do without metaphysics of any kind. Your conjecture about *my* thinking can only be meaningful to yourself because of *your own* introspective activity. You too have no other, *more direct* source of knowledge about me. That is why the analogy seems irresistible. But do not despair; your introspection need not be purely subjective. Others may be in the same boat. There may be some inter-subjective truth.

Instead of rejecting the analogy, one could acknowledge it openly, build on it and give a public account of each step in the 'controlled thought experiments' about to be conducted. This is probably the essence of a scientific approach and just by going public, the observer takes upon himself the special role of detached, privileged and self-consciously 'inertial' observer. Not only can he then publicly profess the resulting knowledge, but also legitimately use intro-spection and combine it with indirectly but systematically gathered evidence. Provided that 'reasons are given' publicly, we may avoid confounding our own introspection with that of our subjects, in this case the reasoners.

Spontaneity, premeditation and self-confidence

The three most striking insights about people being reasonable are of sponta-neity, timing and the confidence with which it is done. Take *spontaneity*: the sheer possibility that an inner activity could also start inside, be generated and initiated internally, needs emphasizing in this day and age when everything mental is still too often regarded as the mere capacity to react and respond, always needing outside provocation or stimulation. Spontaneous reasoning must be possible. But, whatever may go on inside, *reasonable* action is not accidental and no run-off of a fixed pre-designed programme sub-routine. To assume automaticity, accident, or even purely statistical probability will not do. That would describe something other than reasoning.

Yet, in the school of life, there are no special courses in reasonableness in which one has to qualify before practising the craft. One is supposed to pick that up as one goes along. True, some people are better at it than others, but what exactly that means is far from clear. Parents do not seem to convey the knowledge to their children explicitly. Yet, after a certain age, one is expected to perform reasoned thinking without any external prompting. It seems then that reasoning is a deeply-laid practical skill which, once it has emerged, may be almost infinitely perfected. If spontaneity is true, we may regard the person as basically an heuristic operator, a seeker after, rather than a mere receiver of information. The spontaneous quality of the reasoning kind of thinking must be noted because this would open a lead to acceptance of the burdens and pleasures of true autonomy.

Timing: Even more striking is the insight into *fore*thought, that reasoning deals with events that have not yet actually happened. Is it an inner 'manipulation' of something? How can one act upon something that *is not yet*, does not yet exist? Some internal activity must have occurred before an act becomes overt as a movement, whether of hand, lips or eye. One might say, the reasoning process involves a rehearsal of real-life scenes or events on an internal theatre stage before the curtain goes up and before the public can witness the accomplished fact. For the outsider, that is the effect of reasoning. How else could a skilled act be successfully carried out and repeatedly improved, as is well known? But, since there exists no inner stage and no manikins perform on it, how does the process actually work? While he reasons, the person has not yet attained a state in which he could possibly confirm or disconfirm to himself or others, whether the internal scene matches outer actuality – 'reality', whatever that may be. That can only be done by acting out one's thoughts, making them overt and public. But then it is too late, the process is over. This is the point of the present formulation: The 'internal stage' hypothesis drastically changes normal 'outer' relations of events in space and time. By holding space constant, as it were, the arrow of time is arrested or diverted, though not reversed. In the course of the internal performance of a mental operation, the reasoner controls the external reality to come. At the very least, he is thereby enabled to influence materially his own future and that of others. In this sense, *reason conquers time*.

The scientific literature so far seems to evade this issue of an inner activity and our knowing of it; any analogies used make it even more mystical, it is in limbo. Psycho-physiologists leave the question of whether something can exist inside to the philosophers who, in their turn, leave the detail of how the necessary muscle activity is produced to the psycho-physiologists. It can, nevertheless, be said that, unless an action is somehow overt and publicly accessible to others, including the privileged observer, that action has no *outer* reality. As to *inner* reality, it has been demonstrated experimentally at least 40 years ago that thought and implicit activity can occur simultaneously, when muscle action potentials were recorded during imaginary voluntary movements (Humphrey, 1948). Of course, the voluntary imaginary aspect could only be verified by the active introspection of the subjects.

As long as one holds on to the view that actions produce something real, and so *create an outer reality*, one is driven to assume that being reasonable is when one performs operations upon something *before* it has become outer reality. Yet, reasoning cannot be gainsaid as an *inner* reality. The reasoning must have been done by the person, therefore it must have some kind of inner existence. So we have strong prima facie grounds for saying that reason exists and is real, even if its objects and objectives are in the future. Before saying more about timing, we must deal with our confidence.

Self-confidence and decisiveness

It was I who has just been thinking and reasoning and who decided to write these lines. They are the effects of my Self, my thinking Ego in pure self-consciousness – pure because ultimately not contingent on outer experience. This introspective experience gives rise to the third insight to mention here: the confidence with which reasoning is carried out. Considering just how complex the process must be, this self-assurance is nothing short of remarkable. It reveals itself in unself-conscious acts, particularly in skilled ones, subordinate to the 'higher-level' reasoning, when one must launch out from a secure base of self-reference onto the world at large. Well removed from awareness though it may be, this confidence is no less real. The sureness of walking, gripping and performing other so-called simple tasks expresses it and, as we go up the scale in complexity of mental process, the outcomes of our actions become less and less predictable with each added dimension; yet it is not complexity as such which affects the confidence of the reasoner. Only when he suddenly becomes aware of his *un*self-consciousness, does he hesitate. In cycling he loses his balance, in reasoning he may doubt the evidence of his senses; confidence in his own inner process does not normally seem to diminish. If anything, it is consciousness of the ongoing process which interferes with the skill.

The reasoner's confidence is the observer's witness to the existence of inner events without which reasoning would be inexplicable. That confidence probably expresses the *degree of control* the person has over his thought processes. For, if there is a difference between thinking and reasoning, it is that the latter is controlling the flow of thoughts.

Internal representations

We said above that to reason is to conquer the future. The future *world* can only be conquered, if what is thought to be the real world is accurately replicated inside the head, so that mental operations can be performed upon its multifarious representations and simulations of events. These will be the normal *controlled thought experiments* conducted by everyone. The concept of internal representation has a long philosophical history. In German philosophy, in particular immediately following Kant, a specific basic faculty, *Vorstellungsvermögen*, is posited. It is, however, only recently that the term makes its appearance in the Anglo-Saxon literature of cognitive psychology. Reifications in physiology were called 'engrams' 'cell-assemblies', but nowadays computer analogies are used. What follows may be seen as a tentative contribution to the description of representations and an exploration of bases for new theories.

The task can be delimited provisionally by listing, in the light of the foregoing considerations and current research, what these representations are NOT LIKELY TO BE:

(a) They do not directly represent specific objects as perceived. I base this on neuro-psychological evidence that the perceptual process is constructive, interactive and indirect, rather than merely reproductive, passive and direct. Each object of the outer world is de-composed into its features, represented and on recall has to be re-assembled.

(b) Representations are not 'memory' in the sense of a static, passive, rigidly fixed record of the past, but a flexible, continuously maintained and refreshed, fluid process of varying intensity.

(c) The representations are not a random collection of fixed action programmes or plans. Since no two situations encountered are exactly alike, yet are efficiently dealt with in the outcome, reasoning must have been able to modify whatever is preserved from the past.

What internal representations should be can be conjectured thus:

(a) In the first place, they must present the Self to itself and in the second, its relations to the rest of the world.

(b) They represent rudiments and principles of action plans, of images and symbols, including possible projected courses of action, rather than programmes.

(c) They represent, moreover, potential outcomes, results or *consequences* of projected courses of actions.

(d) Representations must be open to selection. There must be a 'choice facility'.

(e) Representations, particularly those from external perception, probably have an emotional charge. Such a charge constitutes a subjective valuation for the person. The 'valuation facility' enables choices to be made before the action is executed.

Self representation

To produce the known results, the Self in all its manifold manifestations must be represented internally. We can see how it works if we combine observation with indirect knowledge. We observe an infant lying in its cot, repeatedly touching first the side of the cot and then its own toes. Even with our limited knowledge of neuro-physiology, we know that there must necessarily be a difference between perception of 'not-self' and self. In the former case, one set of sensory signals is received in the infant's brain: from the fingers only. In the latter case, there are two simultaneous sets: from its fingers and from its toes. The simplest generalisation of this experience is: One sensation = not-self, two sensations = self.

Needless to say, representation of the self must be a forbiddingly complex process, since it must also be involved in all perception of the outer world. Not only the somato-sensory cortex, but whole columnar tracts of the nervous system are probably devoted to this function into which much theorizing has gone. (cf. Granit's gamma efferent system, von Holst's re-afference principle and Miller, Galanter and Pribram's TOTE). The idea of 'body-image' is not nearly adequate enough. Introspection easily confirms how extremely accurate these systems work and how they are a deep and dynamic form of 'memory', e.g. when we step on an escalator that is not working as we normally expect it. Even when we walk or run or jump, we push the ground away from our Selves with our feet, so that this system must also be fully conversant with applications of Newton's third law of equal and opposite action.

At least as complex and no less accurate must be the internal representations of the outer world. This must be the totality of personal experience contained within the skull, not some feeble, particulate traces. Here I would counsel the views taken by Bartlett in his 'Remembering' (1932) and 'Thinking' (1958), rather than those of most present day psychologists. Their line of approach, unfortunately, does not seem to help in solving the problems of spontaneity and autonomy, but goes further and further into the mystical realms of 'mechanisms'.

There is, for instance a practical problem of ultra-short-term memory, which is almost completely ignored by present day empiricists, experimentalists and cognitivists. At the micro-level of all receptor functioning, perception is found to be almost certainly not a continuous but an intermittent process. This is particularly the case with rapid eye movements – the REMs now given prominence in sleep research, but known since the 1950s from Ditchburn's work (Ditchburn, 1963). Yet we are convinced that the world outside is one continuous whole. I know of no micro-study attempting to explain how and at what level that continuity so apparent to us all is achieved. There is a case for suggesting that a micro-memory exists between two eyeball positions and the resulting unstable retinal images between saccadic and other tremorous movements. I would argue that these memories are representatives of self-generated processes and therefore fundamentally important psychological events, which help to maintain the person's integrity and autonomy in an uncertain and everchanging world.

The phenomena of 'internal rehearsal' require a much more dynamic explanation of events than the usual views of memory as of passively stored objects. It is more likely a comprehensive, live information system, actively clamouring to up-date itself continuously from the outside. There are no objects in it at all, no pawns or other chess figures. Neither are there any levers, cogwheels or spring coils to do the pulling and pushing. The use of the word 'mechanism' is only a convenience of researchers, all of whom know full well that whatever mechanics rule inside, they are more like quantum mechanics than Cartesian ones. Neither are the 'intra-cranial' operations to be understood as surgical interventions; the term is used to express activity rather than passivity of process. Memory, as a process, must be global rather than cellular. It must necessarily aggregate dispersed elements, though not in a statistically random manner. Reasoning is too precise and successful to be no more than a trial-and-error business. Anyhow, can error be defined without specifying what a success is? I think not. There must be a notion of a target, aim or purpose, however obscure, to infuse overt action into the internal process.

In accord with the idea of spontaneity, we can regard the person as a basically heuristic operator. As such he is continuously, implicitly questioning the outside world. He abstracts from it certain features and creates his *in*-formation as representations. There is no reason why these should all be 'encoded' in verbal form. Dennett (1979, p. 161) suggests information could be encoded in one of three forms: propositionally, imagistically or analogically. My guess, based on evidence from study of skills, is that by far the largest part of *in*-formation is in the analogical mode, or as Bruner (1980) called it, enactively, in interoceptive, somatosensory and kinaesthetic modes.

The implicit, often non-verbal questions the heuristic operator puts to himself – if they were in words – might be: Is this a table? Is it round? How many legs? Is it really there? Do I like it? The material represented, with which the percept is compared, is, I propose, the same as the elementary units discovered as neural processes of abstraction by feature analysis, by Hubel and Wiesel, the 1981 Nobel Laureates in Medicine. If indeed such elementary units are representative of the vast variety of attributes of objects experienced, the mind is able to form implicit propositions by relating them to each other and to the Self. Just because this is a *process of relating*, no more specific, single and exact cellular locus can be found in the head of objects and events. Representations of external experience are, however, probably place-marked and time-marked. Recent work on 'cognitive maps in the brain' (O'Keefe and Nadel, 1979) found individual cells which fire if and only if the subject is in one specific place.

Marking the time

One often overlooked aspect of time in reasoning is its tense (but for a philosophical treatment see Mellor, 1981). Every sentence, implicit or spoken, has a time-marker, we say 'it has a tense': I was at work. I am walking home. I shall arrive at the doorstep. I shall then have been away all day. I shall have no food for eight hours and I shall therefore be hungry. Hadn't I better bring home some food? These are truly propositions to oneself. In accord with the arguments that reason conquers time, the process must convert the representations of the past into conjectures of future situations. They are then not just imaginary re-constructions, but must beyond that satisfy the commonly held view that reason projects the possible consequences of particular courses of action. (The real skill of reason is to separate the possible from the impossible, the realistic from the unrealistic consequences of one's actions. Based on the observations of D. Ingvar (1979), the conjecture is that this 'consequential ruminating' process takes place in the frontal lobes. Ingvar showed by a very elegant technique that normal cerebral activity during rest tended toward 'hyperfrontal' distribution. This kind of reasoning is an 'optional' extension of mental activity and that explains why some patients can live modest lives without frontal lobes.) In other words, what must be represented, is not only the future, but also the situation *after* the future. (The grammatical term for this tense is paulo-post-future.) On this and other grounds it can be said that most, if not all, internal representations are in the future tense or at least capable of being 'futurised' and are looking back at the present. Insofar as these representations are abstracted from the world outside and time is arrested, they make the world 'portable' for the person and so add to the independence of existence.

Choice and determinacy of behaviour

If only one course of action were possible in each situation, the person would be an automaton. We must assume that humans always have at least two such courses open: to do it or not to do it. Once we observe an action accomplished, we must necesssarily conclude that an *internal choice* has taken place, however unconsciously. This assertion presents a challenge to the fundamental notion of determinacy of events. The full reach of this challenge cannot be argued here and only a brief attempt at resolving the problem can be outlined.

At this point, attention is drawn to the work of Grete Henry-Herrmann and Gustav Heckmann, because the considerations of the tense in which inner representations are held, readily merge with their views on determinants of behaviour and extend them. Henry-Herrmann (1953) draws a clear and fundamental distinction between the causal interpretation of natural events and another category, specifically applicable to thoughtful acts and behaviour (*besonnene Handlungen und Verhalten*), which cannot be fully accounted for by causal description: acts which are determined by the quality of rightness of wrongness of the supporting argument, rather than by its quantitative strength or weakness.

Heckmann reinforces this distinction. Whereas in nature the strongest force prevails, in thoughtful behaviour it is reasons which determine action.

> This determinacy is at stake in heuristic thought, in the consciously responsible decision-making in conflict situations, in the search for the means suitable to the attainment of an aim striven for, in the decision on preferability of one among competing aims (1981, 64) (Transl. P.B.).

Let us separate for the moment determinacy from causality. In the past, scientists, including behavioural scientists, proceeded from the assumption that every event in nature is fully determined by the circumstances of the immediately antecedent events. (This assumption has now been found wanting at the extreme micro-and macro-levels of physical processes and it is thought that events there are indeterminate.) This is one of the dichotomous situations which put to test my resolve to apply critique. One can, I suggest, argue that events may be determinate or indeterminate, but that there may also be two kinds of determinacy, one of events in the world outside us and another of those within us. Whether the outer world of the universe of inanimate matter is 'really' determinate or not, can be subject to theories and experiments but we must ultimately rely upon speculation for an answer.

On the determinacy of the inner world of humans, we can have a little more direct knowledge, because we can ask them as well as observe and introspect. We then find a certain regularity and lawfulness of behaviour in two respects, both utilising temporal characteristics of their inner representations: one links behaviour *to the past*, the other to the future. On the understanding that every event has antecedents in time, we call an observed event an effect and so link it with a specific *cause* in the past. Causal determination is then the establishment of antecedents affecting or creating consequences. But we can also be shown to possess the capacity of representing to ourselves the future consequences of putative actions. If these self-representations were to determine any subsequent, overt behaviour, this would not be causal but *purposive* determination.

One can, of course, argue that the one category does not necessarily exclude the other from applying to a particular case. One's behaviour can be determined both by past experience *and* by future 'intention'. I have, however, regarded that problem as one of the nature of the internal representations and the tense in which they are held, and believe they are – for all practical purposes – in paulo-post-future. In either case, one thing seems clear: for the more complete explanation of behaviour, both the causal and the purposive categories are needed.

This three-fold division of a *categorical 'Moment of Determination'* has very fruitful implications. As soon as we are once able to overcome the monopoly of causality of determination, we are both free to choose and burdened with choices. Whatever the process of making choices, the critical moment in time must be *before* execution of the act. Otherwise, giving an account would be *post hoc* speculation, called rationalisation, rather than reason proper. For the reasoner, it is a decision procedure, whatever we as his observers may ratiocinate.

Preferences and subjective values

Given that the reasoner makes choices, they probably consist of the fairly well understood processes of comparing information, in this case outcomes, with each other and 'preferring' one. By what criteria is a particular course of action, a behaviour, preferred? No instant value system can be provided here, but I argue elsewhere that the process is relational, rather than absolute, and that subjective likes and dislikes may form the first detectable basis of valuations in actual experience. I see these as 'crypto-valuations' because they need not be conscious and deliberate. Evidence from extensive discussions, interviews and introspection leads me to conjecture the process thus: To begin with, a striving-like feeling gains in clarity step by step when adumbrated by reflection. The most frequent occasions for introspective reports derive from intense feelings of unease, of dissatisfaction with a given situation, of 'indignation'. 'This is an untruth.' 'This is an injustice.' 'This is an indignity.' Often enough, no conceptual clarity is reached but only so-called intuitive understanding. What are at first subjective valuations can later be generalised and modified, even turned into their opposites and made progressively more inclusive of other persons until they are 'inter-subjective'.

Finally, to appreciate the place of emotion in being reasonable, we must return to the outset of this inquiry: the contrast of calculating rationality with emotional impulsivity. I believe there is general consent that our target lies somewhere between the two. The active aspect of being reasonable must necessarily have required some impulsion which mere uncontrolled thinking alone cannot provide. We know of no other psychological 'motivating' process, phenomenon or mechanism which can bring about an equivalent effect. Values must be used in choices and will, as emotions, lead to expressive action. Values and emotions yoked together produce the result we observe. In this way some emotions are controlled by thinking.

As they deliver the criteria for choice, values are an integral part of reasoning. What these values are, whether the 'highest values' are objectively valid or only subjectively so, and whether they are *true*, is not under discussion at the moment. If one is *de facto* chosen as such, the description of the event in inner nature is satisfied. This then is the operational description offered here: 'Being reasonable is the practice of choosing – before acting – from among available courses of action the one with the most highly valued consequences'.

Science and critique

In the above, all too brief, outline two far reaching extensions to present scientific studies are implied: one is the necessary and conscious inclusion of 'metaphysics' in theory building generally, the other is that purposive explanation of behaviour necessarily leads to a renewed effort to put *social interaction* on a scientific footing. Both flow from a re-assertion of the value of critique as practical, systematic procedure, not merely confined to purely speculative philosophy.

The critical method applied to science must proceed from facts of observation. This is not denied here for one moment. But far too often, no account is given why just that fact and not another is used as a starting point. Yet everything for a theory must depend on that choice. Popper argued very eloquently for a trial and error process in science. However, it is also a fact that many eminent scientists have turned to metaphysics for answers, e.g. Whitehead, Schrödinger, Max Born, Niels Bohr. Recognising then that metaphysics must inevitably be involved, it should become good practice for the scientist to search critically for the 'metaphysical' assumptions underlying his own judgements. Instead of capricious choice, a more conscious selection of facts may be possible. How effective this is would depend in part on the value system held. This procedure does not do away with trial and error assessment, but the trial is by reason rather than by chance. Moreover, even if metaphysical, the criteria of rejection as error or acceptance as 'probably true' are then explicit. As I hinted above, without that search for abstract criteria, how else could one tell what represents an error in any trial?

My demand for a science using metaphysics explicitly and critically is especially aimed at diminishing the ever present danger of a vacuous *mysticism* taking over from an equally empty *empiricism*.

A science of social interaction

My attempt at a purposive explanation of human behaviour is, I have no doubt, in the tradition of Kant, Fries, Apelt and Nelson, even though they did not state the problem in this form. In Nelson's time, academic attention just began to turn to behaviourism as an extreme empirical explanation. However, 65 years later, this is shown to have been a passing preoccupation which has run its course. While cognitive psychology has in the last decade freed itself from that imprisonment, it must now complete the process and temper scientific experience with realistic acknowledgement that some aspects of our knowledge of matter *exist only in our thoughts*.

The term behaviour can only be meaningfully applied to interaction between sentient beings, as distinct from inanimate matter. With the latter there are no *inter*-actions since, strictly speaking, it has no will of its own and it can only respond in a manner either indeterminate or, if an artefact, determined by its designer. The participants in interaction, by contrast, as it were, are jointly and severally engaged in natural processes of their inner nature and so subject to a science *sui generis*. Whether such a science can produce valid predictions or only descriptions, remains open. But it may validly occupy itself with detecting lawfulness and regularity of associated events.

One of the regular associations I hope to have recognised is the yoke connecting values with emotions. In my view, this relation is more important as an explanation of the natural process of reasoning than a prior insistence that values must have an objective foundation. It is surely valid to distinguish between form and contents of value judgements. Then, whatever the *contents* of such a judgement – whether openly stated or crypto-conscious – criteria for recognising it can be set up. Regardless of dogmatic assumptions about origin and contents, values may be subjective or objective, if coupled with emotions they influence social actions. As the person values a purpose, so he/she will choose to act. Behavioural scientists can legitimately investigate these processes.

Since the future-determined category is the quest for understanding purposes, we open to science not only social propositions as interaction between two beings, we also re-open the scientific study of ethical (or unethical) behaviour. For this extension I take as a working hypothesis that all interaction is social, and it becomes especially ethical when the interests of two participants clash. Ethics is then the scientific study of conflicts of interest and the resolution of moral conflict as a series of processes in inner nature becomes a new and worthwhile task for scientists.

References

Bruner, J. S., 1980, *Beyond the Information Given*, London: Allen & Unwin.

Dennett, D. C., 1979, *Brainstorms*, Brighton: Harvester Press.

Ditchburn, R. W., 1963, Information and Control in the Visual System, *Nature*, **198**, 630.

Granit, R., 1962, *Receptors and Sensory Perception*, Yale: University Press.

Heckmann, G. 1981, *Das sokratische Gespräch. Erfahrungen in philophischen Hochschulseminaren*, Hannover: Hermann Schroedel.

Henry-Hermann, G. 1953, Die Ueberwindung des Zufalls. Kritische Betrachtung zu Leonard Nelsons Begründung der Ethik als Wissenschaft, In Specht, M. und Eichler, W. (hrsg.) Leonard Nelson Zum Gedächtnis. Frankfurt/M. Oeffentliches Leben.

Holst, E. von, 1973, *The Behavioural Physiology of Animals and Man*, London: Methuen.

Humphrey, G., 1948, *Directed Thinking*, NY: Dodd Mead.

Ingvar, D. H., 1979, 'Hyperfrontal' distribution of the cerebral grey matter flow in resting wakefulness. *Acta neurol. Scandinav.* **60**, 12–25.

Mellor, D. H., 1981, *Real Time*, CUP.

Miller, G., Gallanter, E. and Pribram, K. H., 1960, *Plans and the Structure of Behaviour*, New York: Holt.

O'Keefe, J. and Nadel, L., 1979, *The hippocampus as a cognitive map*, OUP.

Chapter 16

Process control operators as responsible persons

Paul Branton

Invited Paper for Symposium on Human Reliability in the Process Control Centre, Institution of Chemical Engineers, Manchester, April 1983

Abstract: Monotonous work is rarely considered a source of stress. Yet, when routine tasks with responsibility for safety are carried out in a monotonous environment, operators experience considerable stress and fatigue. Behavioural observations show many situations of low arousal states, with introspective episodes of disorientation and loss of time sense. Sudden emergence from such states is accompanied by 'mini-panics'. Experiments are reported showing natural daytime fluctuations in arousal of about 100 min duration, with tendency to daydream if self-activity is minimal and environment impoverished. These phases are similar to night-time REM sleep stages.

Introduction

The premise of this paper is that the process control operator is in fact a responsible person even if he is not always treated consistently as such. If a man is to be responsible, he must be given the means for this and is not to be regarded as simply an automaton, a cause/effect machine. Non-biologists may be excused for overlooking the point but they should realize that physiologists and psychologists, in their descriptions of human behaviour, may unwittingly mislead us all into confounding analogy with the real thing. In theoretical explanations, they still speak of 'brainmechanisms', 'automatic reflexes' and 'stimulus/response sets', as if no choice intervened between input and output. Alternative explanations of the nature of human cognition and action are just emerging from controversy. We are looking now at a 'person' and deal with probabilities, potentialities, capabilities and faculties. Our understanding is, as yet, uncertain and at best only statistically predictive. In that respect it is not so different from the knowledge the nuclear physicist has of his raw material.

We shall concentrate here on the contribution the individual operator can make to the avoidance of 'human error', insofar as he can do so himself. For, although he is far too often called upon to compensate for the equipment designers' lack of understanding of human functions, he cannot, strictly speaking, be held responsible for design 'errors'. Our main concern now is the study of the individual, to draw attention to recent research results from a wide range of disciplines. Only then can a managerial and organisational responsibility for the provision of reliable and safe working conditions be usefully advanced.

No one denies that the presence of human operators in a system inevitably turns it into an *open-loop* system, subject to error, however small the probability for even the most conscientious person. Our position on safety research and accident prevention, based on evidence from many field studies, is this: *Reported* accidents rarely have a single cause, but are the results of a number of multiple-cause events, each with very low probability, coinciding at one point in time and space. Just because accidents are so relatively rare, one cannot expect to learn much from experiencing them. In the absence of sufficiently substantial knowledge, it is therefore necessary to search for classes of rare events, in particular certain types of human actions, which are likely to involve risk, with the aim of anticipating their 'unforeseen', undesirable consequences. In desperate circumstances, it can be said to be a duty to speculate about the future and not merely to rely on past experience.

Errors and 'harmful' stress

It is often assumed that errors are more readily committed when a person is under stress from too much work. But this does not seem to be the case in all process control situations. On the contrary, a large number of them are more a case of keeping awake than of keeping calm. Not mental or physical *over*load, but boredom, as purely mental *under*load, is the problem. A certain degree of arousal is actually desirable in the circumstances. But when does stress become 'harmful'?

The evidence for this triggering process, turning a wanted state of arousal into one that is perceived to be harmful, comes from extensive behavioural records and the related intensive self-reports, often by very articulate subjects. It emerges that, at certain times and in certain monotonous conditions, moments occur when the operator spontaneously increases his activity considerably. Two reasons are given: One is an express urge to seek sensory input, a need for stimulation. If the work does not provide this input, the individual seeks or produces it himself, even if only by getting up and walking around, provided the job permits this. This interpretation of observed behaviour tallies with what is known about living organisms in general, and modern man in particular: There is an irrepressible, spontaneous need for sensory inputs. (There seems to be a lesson for environmental designers here, to which we shall return later.)

Mind wanderings

The other reason for sudden spurts of activity is that operators report that they recall having struggled with drowsiness and realised that their 'minds had wandered' from the task. They knew that there was a time gap during which their mental activity had slipped out of control but are quite uncertain about the duration of that gap. This experience of itself does not produce harmful stress. That does, however, arise as soon as the person becomes aware of the possibility that a mishap may have occurred during the moment of lapsed attention. In other words, it is the awareness of personal responsibility which seems to generate the real stress. If, in addition, the operator is *uncertain*, either about the possible mishap or about his capacity to resolve the urgent problem, anxiety increases. Even though these are experienced operators, they do not seem to be able to anticipate or prevent some of these lapses.

Here are thus two pointers to the nature of the harmfulness of stress. The loss of *time sense* connected with absence of mind and the *degree of uncertainty* in the situation. Both are concepts amenable to further research. The immediate practical point to make is that if appropriate information and training can be pre-arranged and time pressure reduced, harmful effects become less likely and uncertainty more bearable. (In passing, the often used explanation of stress as the 'fight or flight' response, is shown to be wholly inapplicable in the present context. The responsible control room operator has neither anyone to fight, nor can he simply withdraw from the scene.)

Emotions and interest

The self-reports leave no doubt that even uneventful shifts in control rooms and similar work places are subjectively very fatiguing and that the sudden 'mini-panics' can be emotionally exhausting experiences. But since there is almost no physical energy transformation involved, the very real feeling of exhaustion would seem to be the result of emotional commitment to the task. While the relation of emotions to activation is known, it is not yet well understood how alertness can be maintained or enhanced deliberately. We believe it is necessary to introduce the concepts of *interest* and *purposivity* to explain the processes of alertness and selective attention.

Bodily functions and levels of consciousness

We now come to a somewhat speculative interpretation of factual evidence, which nevertheless has face validity. The evidence comes from two independent areas, sleep research and production engineering. Field observations of repetitive manual work (in connection with accident prevention)[1] and in vigilance-type tasks[2] have shown periodically recurring fluctuations in alertness and performance. The cycles recurred at intervals of about 90–100 min. As this cycle time is the same as the now well known REM sleep stage cycles, and REM has been reliably associated with dreaming, a close look for similarities between the two situations was taken. The findings of sleep researchers are that, in the right circumstances of enforced inactivity throughout the waking day, almost everyone is subject to such changes in arousal at about 90–100 min intervals. They are regularly reported to be accompanied by brief spells of daydreaming.

Of course, both perceptual intake and body movement is suppressed during sleep. But even if one is not asleep, yet in a well-supported posture, whether horizontal or seated, the anti-gravity muscle action control is 'switched off'. If, in addition, the eyelids droop and visual input is suppressed or diminished, the person enters into twilight states in which the balance between attention to the environment and to the self becomes unstable.

These transitional (hypnagogic and hypnapompic) states are frequently associated with highly original and creative thoughts and sometimes with hallucinations, in other words, with fantasies. A connection is thus suggested between secure suspension of the body and reduced information intake on the one hand, and enhanced abstract mental activity on the other. Conversely, it is arguable that two factors which normally prevent one from being conscious of those states of diminished external awareness, both in daytime and during the transitions to sleep, are the active search for stimulation, particularly visual, and anti-gravity muscular work. In short, daydreams are less likely to occur if one keeps oneself upright and actively occupies the eyes and ears and the limbs.

The urge to escape from situational uncertainty in a compulsively restricted environment, like an isolated control room, can be seen to evoke the response of the operator of creating his own inputs. There is a risk that this quasi-information differs from the real state of affairs as indicated by the instruments around him. If the environment is impoverished by soundproofing or monotonously invariant climate, albeit well-intentioned, the 'real' world outside is more remote and uncertain, an invitation for the mind to wander.

Stress and social consequences

If the operator is responsible, he is a *self-determined system*. What he determines is his purposes and he can do this more or less at any level of present awareness. In fact, he obviously can to some extent consciously override his own bodily rhythms – provided his purposes are strong enough. In this, his determinant is the *consciousness of the consequences of his own actions before they are carried out*. That is the essence of his rationality, implied by investing him with responsibility. It can be argued that the process of determination is general, but that what the operator regarded as 'consequences' may vary. They may be real or imaginary, according to the momentary balance of awareness of the inner or the outer world. The operator can and must be made aware of the possibility of illusory perception and imaginative construction. One of the factors influencing him in this is, no doubt, the social one. The real stress which operators reported, as mentioned above, was almost always related to their view of the implications of possible mishaps for other persons, in or outside the control room. The presence of others in the room can thus have two effects worth noting. They can be referred to in case of uncertainty about the real state of affairs and they can add to the social motivation and involvement in the task.

Self-management

The above findings and speculations are reported here in the hope that knowledge of them can inform the responsible operator's self-conduct and self-management. Much anecdotal evidence of such self-management is found among shiftworkers in regard to their sleep and food habits, or in the way pilots cope with jet lag. Systematic development of positive techniques of stress and fatigue prevention would be one avenue into which managerial effort could be directed. Other measures to enhance his own alertness the individual could adopt may be descibed as follows:

Individual

1. Consciously self-induced changes of posture (about one per minute) and occasional movements of the whole body, especially when low on the arousal curve of 100 min.
2. Deep breathing for transport of oxygen to the brain.
3. Occasional check of time on task – rather than of clock time – to establish position on arousal curve.
4. Intake of glucose/fructose or caffeine.
5. Cooling of head, ears, wrists, ankles to counteract too warm or stale climate.

Organisational

1. Active programme of stress/fatigue prevention with periodical review built into the system.
2. Enironmental design
   (a) avoidance of rhythmical 'lulling' stimulation in sight and sound,
   (b) introducing mild random climatic variations, frequent air changes and humidity/ionisation balance.
3. Provision of secondary activities not competing with main task. Operations to be designed for standing as well as sitting, and for some walking about.
4. Job enrichment, alternation and rotation.
5. Creation of positive interest in task.

References

1. Branton, P., 1970, *Int. J. Prod. Res.*, **8**, 93–107
2. Branton, P., and Oborne, D., 1979, A behavioural study of anaesthetists at work, in *Research in Psychology and Medicine* eds. Oborne, D., Gruneberg, M, and Eisler, J. R. London: Academic Press.

Chapter 17

Critique and explanation in psychology — or how to overcome philosophobia*

Paul Branton

Unpublished paper, 1983

Although I never knew Sir Frederick Bartlett personally, his works and the reminiscences of his pupils convince me that he must have been a truly Socratic teacher and practitioner of the art of midwifery of psychological thought. His refusal to dogmatize comes over strongly and so does the heuristic aspect of his style. No wonder he gathered around him some very fertile brains, Kenneth Craik among them, and encouraged them freely in their search for ever deeper explanations of behaviour. In a little book, Craik (1943) made two brief remarks about a strange and difficult concept, hylology, which have, so to speak, rankled in me ever since I read them 24 years ago in my undergraduate days. I have now traced his meaning through to a long line of philosophers of mind, which links Craik and Bartlett directly with Fries, Kant, Descartes and then straight through with Socrates himself, circumnavigating Aristotle and the dogmatic parts of Plato on the way.

The ingredient I found their works to have in common is critique, a method of approach so subtle that it is easily misunderstood by the impatient. It seems to me particularly suited for application to psychological research, since it takes its point of departure from that which is given to the senses of the observer, 'naturalistic' study of human behaviour, as it used to be called. Now, the more apt but equally question-begging term is 'ecological'. I am convinced that the advocates of this approach do not want to restrict themselves to the physical environment, and rightly so. They really mean to take a holistic view, to treat events in their full *context*. The question then becomes, 'Where does it end?' For instance, I cannot know everything before deciding to act, yet decide I must.

Critique can be contrasted with the usually practised dogmatic method when it comes to choice of research topics and criteria. If this raises the discussion to the level of meta-psychology, so be it! Whereas the initial choice of the questions

---

* Fear of philosophizing, reluctance to search for first principles

177

to which answers are sought presents little problem to the dogmatist when he sets up his hypotheses to fit his theories, the discerning and heuristic researcher turns to critique and asks for the – as yet – hidden meaning of what he observes. Hidden, because no amount of 'hard data' can of themselves alone bring the insight into the connective ideas and relational abstractions between observed events which is needed to uncover new knowledge. I have come to think that Craik, and Bartlett, actually practised this tracing back of a thought about experience to its underlying presuppositions. It is critique, not ordinary induction, because it deals with presuppositions rather than factual events. It is *hylological* critique because the results are the observers' abstractions from their observations of concrete persons, the products of their own minds rather than of the sensory input into them.

Hylology can be contrasted to morphology. Morphology, as study of the form of organisms, deals with the appearance of matter in space (including things and persons) as given to perception, their '*Gestalt*' in the original meaning of that term. Hylology is what we think about those things, our own contribution to the process of internalization. If *Morphé* is about discrete, perceptible, events, *Hylé* is about the interstices between events, the connective tissue, so to speak. Both, it must be stressed, study the very same object or subject person, but from two disparate sides. (Let monists and dualists go on arguing their barren dogmas among themselves. Neither of them can gainsay that causality, necessity, universality and a number of other concepts are thought by us but not perceived directly with any sense organ. Sooner or later, psychological theory will have to explain our possession of these concepts.)

Critique is not theory. The relation of critique to explanation is that the former first seeks to establish what it is that needs explaining, whereas explanation consists of theories in the form of conjectures and abstractions about the phenomenon being studied. The first thing to be said for critique is that it encourages the habitual focusing on what questions are to be asked before rushing out of the lab again for further evidence. By concentrating on the actuality of sensory experience before formulating questions, one remains on firm ground in the choice of topic, instead of pursuing one's own or someone else's whims and fantasies. I have found this discipline most valuable in the practice of industrial consultancy where the diagnostic function is often more important than problem solving. It also confirms the burden of caution put on us by philosophers like Popper (1972).

An even more fruitful aspect of critique is that its main question is epistemic. It asks about the *necessary and sufficient knowledge* without which the persons observed could not have behaved as they actually did. What did they know and when? What information did they seek? (Perhaps to be judged by line of gaze?) What, if any, information did they receive and when? By which modality? Above all, what was the sequence of events, the timing? Strict reconstruction of such evidence reveals the crucial gaps in the researcher's own knowledge and thus the targets of his own heuristics. [Putting these questions consistently to myself while researching into a variety of complex skills, I have not been disappointed with new areas opened up to investigation. (See Branton, 1978, 1979, 1983.)] The necessary-and-sufficient clause tries to ensure the parsimony of Occam and that can hardly be a disadvantage.

Some examples of the gaps I found in explanatory theories can be given to illustrate:

1. Internal representations of the world outside: Objects in that world are doubtlessly represented in the head. What form does this take? Two principles, at least, seem to be at work, one of transformation and the other of distribution of elements for economic storage.

    (a) Principle of Transformations: (3D→1D→4D) If mind is in the head, its morphotic side presents itself to the researcher in the binary code of nerve impulses, effectively all-or-nothing events and therefore one-dimensional. But when thoughts emerge into overt behaviour, they have a time-marker attached; which makes them four-dimensional. We must explain how this happens.

    Everything that enters the head must do so via sensory receptors and through the central nervous system. There is as yet no other known way to storage. If this is true, a number of consequences follow. One of them is the strong implication that all perception must be of a self-constructed nature. Perhaps like Richard Gregory's 'Fictions'? (Gregory, 1974) Just as the eye is NOT a camera, so everything else received is subject to transformation into nerve impulses, indubitably binary. Therefore, whatever is represented, and in whatever form this may be, it must be subject to all the vagaries of this transformation into a time-marked item, before it can be uttered again as a report of experience. Forms of representation are discussed below and elsewhere (Branton, 1982).

    (Another consequence of this principle is that it differentiates scientists from mystics and occultists. The onus is on advocates of paranormal events to show, not just what is in the head, but how it got there.)

    (b) Principle of Distributed Storage: If we take the foregoing together with the discoveries of feature analysis of objects by Hubel and Wiesel, and Blakemore, it becomes feasible to regard features as distinctive elements we 'attribute' to objects perceived. It is then no longer necessary to store every single object we encounter separately in the head. Squareness, whiteness, smoothness and all other perceived (morphotic) features of a piece of paper can be stored in widely dispersed locations in the brain. This is really economical, because each attribute can be shared with any other object, analogous (!) in this respect, even if vastly different in any other feature. What calls for (hyletic) explanation is how we bring them together again when we tread the 'final common path' of expressive action, which is always muscular, whether verbal or not. It is possible to say that the object is represented internally only as the 'common meeting place' of all its criterial attributes, when these attributes are called together by some effort on a later occasion.

    As to the forms in which information about the world outside could be stored, Bruner (1964) suggested three modes: enactive, iconic and symbolic. I believe this way of description has heuristic value, because the enactive representation of skills could explain why we have such difficulty in describing with language symbols how to ride a bicycle or other psychomotor facility. These actions may be stored as muscle coordinating command programmes, not normally requiring transformation into language symbols.

2. The gap between thought and action: Most cognitivists seem to assume that an action is 'no sooner thought than done'; skill researchers know differently. Both of them, however, seem to take for granted that cognition is a sufficient condition for

'motor behaviour' to occur. If I stood up and lifted my arm, I would fall on my face – unless I corrected for the shift in my overall centre of gravity. Acting persons have all this in their minds, even if not always consciously. The biomechanical and micro-environmental knowledge about the state of the self in relation to its purposes, an essential need for any behaviour, is all too often taken for granted.

More basically, the fact that every fine detail of an action programme can be overridden deliberately shows that the step from knowing to doing is not an 'automatic' facility. The mere *potentiality of choice* BEFORE action is too easily put aside in morphotic accounts of human behaviour. Psychologists who still believe in old-fashioned conditioning will, no doubt, find this statement difficult to swallow but it will have to be faced some time. Choice comes in everywhere. On perceiving, with a plethora of uncertain information streaming in from all sensory modes, selectivity of attention remains to be explained; at the output end, the fact that we achieve our set purpose in more ways than one only goes to show that a pure myth is still abroad about 'mechanisms', quite unworthy of hard-nosed (morphological) scientists. Craik was very conscious of this and it is why he turned to hylology.

There is an even more weighty reason for hyletic explanation: social and ethical behaviour. That such behaviour exists there can be no doubt. What is moral, what immoral and what a-moral, is for the philosophers to speculate upon; how it is done, IF it is done at all, that is the psychologists' task to explain. But these are topics for another article.

References

Bartlett, F. C., 1958, *Thinking*, London: Allen and Unwin.

Branton, P., 1978, The Train Driver, in Singleton, W. T. (ed.) *The Study of Real Skills I: Analysis of Practical Skills*, Lancaster: MTP Press.

Branton, P., 1979, The Research Scientist, in Singleton, W. T. (ed.) *The Study of Real Skills II: Compliance and Excellence*, Lancaster: MTP Press.

Branton, P., 1982, The Use of Critique in Meta-Psychology. *Ratio*, **24**, 1–13.

Branton, P., 1983, Transport Operators as Responsible Persons. Invited paper presented to CEC Workshop: Human Performance in Transport Operations, 28–30 January, Cologne.

Bruner, J. S., 1964, The Course of Cognitive Growth, *American Psychologist*, **19**, 1–15.

Craik, K. J. W., 1967, *The Nature of Explanation*, Cambridge: University Press.

Gregory, R. L., 1974, Psychology: Towards a science of fiction, *New Society*, 23 March.

Popper, K. R., 1972, *Objective Knowledge*, Oxford: University Press.

Chapter 18

Transport operators as responsible persons in stressful stituations

Paul Branton

In *Breakdown in human adaptation to 'Stress' Towards a disciplinary approach*, Vol. 1, Part 2, (Ed. Wegmann, H. M. 1984), Nijhoff, Dordrecht, Netherlands for Commission of the European Communities.

Abstract. The reliability of a transport operator depends on his ability to control not only the vehicle but also his own states of awareness and alertness. As stress is frequently reported in monotonous situations when mental work load appears to be low, doubts about the appropriateness of current theories arise. Usually, the operator is modelled as an 'engine', as a physiological organism, as information processor or as basically differentiated by personality type. Each of these models conflicts with the reality of transport, where practical necessity treats the human as a responsible person. In field studies, brief episodes of absent-mindedness have been found which normally recur cyclically at about 100 min. intervals, akin to the dream stages of REM sleep. When suddenly emerging from these lapses, 'mini-panics' have been reported. It is thought that the sudden consciousness of having lost control over oneself, whilst being fully responsible for the lives of others, may be the major source of stress in transport. A model of the operator as a goal-directed controller, autonomous, yet liable to natural fluctuations in consciousness, is presented. Some counter-measures to maintain the arousal level are suggested.

Introduction

Reliability is often regarded as the inverse of stress: the less stress a person experiences, the more reliable he is thought to become. This relationship does not hold true in those transport situations where operators are eminently reliable and yet are under very real stress as they are necessarily responsible for the lives of others. No one – and no mechanical or other contrivance – can relieve them entirely of that responsibility. The charge of responsibility tacitly assumes an autonomous, self controlled person, which is right in principle, but which raises problems in the real-life situation because no one can be certain that the operator is always under full, effective self-control. The uncertainty arises from the common knowledge that those body and mental functions which affect awareness and alertness vary and are not wholly controllable.

The thesis of this paper is that any improvement toward self-determined behaviour which can be wrested from the inter-play between man and environment will help to increase overall safety and reliability. The operator as a self-activated being in a mixed man-machine system is, however, not yet well understood. The usual theoretical models of him are not exact enough to bring about significantly greater confidence. They are often internally inconsistent and not wide enough to grasp the generality of real life. Four typical models may be characterized as influencing recent thinking on stress:

1. the 'engine' model – energetic – mechanistic – input/output
2. the 'organic' model – physiological – reflex/response – hormone balance
3. the 'personality differences' model – attitudinal – life style
4. the 'information processing' model – consciously cognitive decisions.

These models will be considered critically in general and then with particular regard to the methods used to study them. Certain practical conclusions are drawn and an alternative model is briefly presented of the operator as a purposive vehicle controller, essentially autonomous but always subject to naturally fluctuating levels of awareness and adaptation.

The concept of mental load

In transport operations man is said to pit himself against physical nature. But engineers and designers 'adapt' that nature to their purposes and so create an artifical environment which stretches the operator to the limits of his own natural, adaptive capacities. Especially in modern vehicles, the operator as driver and pilot neither produces nor expends much physical energy, yet he controls large energy resources very accurately. If he is to be treated as a *reliable controller*, it is appropriate to view him as pitting himself against the limitations within his own nature. The struggle is now mainly a mental one – between keeping calm and keeping awake – a matter of the limits of reasonable self-control.

The complexity of the problem is demonstrated when explanations are sought from a simple analogy taken from the 'engine' model: mental 'load' as a source of stress. Many studies have confirmed the task related differences between the response required of pilots and, say, train drivers. The pilot's task is characterised by relatively short but intense periods of heavy mental load during take-off and touch-down with long periods of relative inaction in between; whereas the train driver must constantly monitor his progress closely and frequently intervene actively throughout the journey. The pilot has too much to do in too little time; the train driver has relatively little to do, other than watch, most of the time. If there is stress, he suffers from underload, where the pilot does so from overload. The difference has received little attention from researchers, probably because of the more dramatic situations of high speed travel. Public transport long distance coach drivers and train drivers have been rather neglected.

Compared with research in other fields, the train driver's task is akin to the less glamorous process controllers in refineries (Bainbridge, 1978) or anaesthetists at work (Branton and Oborne, 1979). As long as all is going well, a low activity level combines with long lag between action and feedback of result. The problem common to these operators is one of knowing whether all is indeed well or not.

The engine model

Even on the engine model alone, the difference between train driver and pilot demands different methods of fact finding, different measurements and different remedies. Direct measurements of energy expenditure by the usual metabolic parameters are not likely to be sensitive enough to bring out that difference. Heart rate measurement is indeed shown to reflect the intuitively judged high mental load of the pilot (Nicholson *et al.*, 1970) and the racing car driver. However, in the low activity, long lag, underload case, heart rate does not reveal directly any response which could be related to stress and the 'mental exhaustion' widely and totally convincingly reported by operators. Could that be because the heart is in the first place a pump and so an energy supplier and only indirectly an index of mental events? There is no need to enter here into a critique of the 'energetic' side of the engine model since this has been done recently and very effectively by Sanders (1981) among others. Unfortunately that does not go far enough. Perhaps the inadequacy of language to express subtle distinctions is at fault; one easily overlooks that any model is at best only an analogy and not the object itself or concept to be described or explained.

Similarly, 'mechanistic' models fail to satisfy most of those who deal with the actual operators day by day. Naturalistic description of events by practitioners, as distinct from theoretical approaches to vehicle guidance, persistently reveals a dynamic mental control activity of such flexibility and immediate adaptability to suddenly changing circumstance that any automatisms and mechanisms become mere figures of speech. In mental functions, nowadays generally called 'cognitive', it is arguable that the term 'mechanism' denotes a mythical entity. There is nothing to be found in the pilot or the driver which functions strictly automatically as in a machine or in a computer. A case can be made that the term 'faculty' would be less misleading.

This is not to say that models have no use at all, just that researchers must not 'adapt' to them and confound them with the real thing. Such is also the case with the cybernetic model, a subset of the engine model. Ever since Ross Ashby, their explanatory power has been used to illustrate, for instance, the generality of the biological tendency towards homeostasis, a concept resorted to in explaining human stress response. The evidence is used to show the ever available capacity to compensate for 'adverse' environmental conditions. When and how the system decides what is adverse and what is not, still remains mythical, or is shifted to a higher level of abstraction. At the level of most frequent usage, to discard these kinds of mechanisms would be a gain in scientific rigour.

Nevertheless, the cybernetic model has the great historical merit that it overcame the static view of the person as just sitting there passively to wait for a stimulus before it responded. The model favours the dynamic view of continual internal process, where a stimulus must 'fight its way into the system'. Furthermore, its considerable heuristic value is that it demonstrated how a weak David of a man can control a disproportionately larger Goliath of a system with an infinitesimally small amount of energy.

Even the sophisticated cybernetic models break down at the practical point at which it becomes obvious to all that the human operator has a mind of his own and is essentially and indubitably an open system. If the system of the human operator is open in real life outside the laboratory, that open-endedness may turn out to be a strength when compared with the rigidity of an automatic, mechanistic device.

The organic model

The physiological phenomena of stress need hardly be elaborated here; they are far better known than the mental events antecedent or concurrent. This assumes that at least one primary source of stress is 'in the mind' and, if true, takes the present discussion deliberately to a stage before that of the more descriptive than explanatory models of Selye, Levi or Frankenhaeuser. The facts established by the physiologists are not made any less true by critique, only their interpretation will be affected. For instance, homeostasis, as expressed in physiological balance is, in fact, often overridden by the kind of social responsibility of pilots and drivers, mentioned at the outset. Really effective remedies for this kind of stress have still to be found. Maybe some effort should now be spent on seeking solutions at a higher level of abstraction in metatheory. As far as the transport operator's case goes, Cannon's 'fight or flight' schema should now be given a decent burial.

On closer analysis physiological models are often a strange mixture. They still present basically a mechanistic, stimulus-response approach, looking for pro-bable external 'causes' of stress where there are obviously multiple causes, some of them internal in origin. Not merely do they look for causal explanation, but tacitly assume a deterministic succession of events directly observable by outsiders. At the same time, physiologists happily accept the possibility of voluntary move-ments and distinguish them from involuntary ones. They thus rush in where psychological angels fear to tread. The possibility of there being an indeterminate or self-determined 'system' at work is thus just as tenable – even if difficult to contemplate because it is unfamiliar. The assumption of indeterminism in human behaviour introduces, no doubt, an element of uncertainty into the discussion, but if physicists and philosophers are able to live with this uncertainty so will human factors researchers.

Personality differences

The evidence of physiology almost inevitably leads to the third typical set of theoretical models: stress as the effect of differences in personality. Whether or not physiological stress syndromes present themselves after prolonged exposure to transport situations would then depend on whether a person is, say, of Type A or B (Friedman and Rosenman, 1974). In this case, a presumed primary causal factor becomes active only when stress becomes 'harmful'. How the body (in its wisdom) 'knows' when that is the case and communicates it to the mind remains an important question for the practitioner.

In the last analysis, one may distinguish a strong and a weak version of the influence of personality upon stress. The strong regards the transport operator's whole personality so entrenched as to be strictly unchangeable. Any attempt at modifying his behaviour would then not only be in vain but also dangerous, as in frequent cases stress appears to result in reversion to basic type. Insights into these basic differences, if true, would only be of practical use if individuals were definitely identifiable, by means of generally acceptable tests, and then only if job selection is feasible on scientific as well as on social grounds. Remembering the controversies over selection by intelligence tests, the prognosis for this approach is not bright.

The weak version of the personality approach to operator stress is supported by the common experience that there are no clear-cut types but an infinitely graded variety of personalities. Moreover, behaviour under stress is not totally invariant within the person from one time to another. If even slight modification of behaviour is possible, it must be attempted for the sake of greater safety and reliability.

One recent development in this area is work on 'locus of control'. It is worthy of special mention as it may lead to a promising growing point. Although recent authors (Rotter, 1976; Lefcourt, 1966) treat the stress problem as one of attitudes or beliefs in the social context of work, they all share the tacit assumption that control over an operator's fate could possibly be in his own hands. The idea of 'perceiving the locus of control within the person' necessarily implies the kind of autonomy pursued here, albeit at a level of actual individual (mental) function, whether or not it is merely imagined. Whether perceived or not, whether believed to be within the person or not, the ultimate control over stress effects on behaviour must be presumed to rest with the person. Who else moves his hand on the joystick or his foot on the brake? Control of an operator's action, not least in the transport context, cannot mean anything other than that a discrete action must be decided and carried out by the individual and no one else. That is the common meaning of individual responsibility in practice. It is applied here explicitly to the initiation of even the smallest muscular movement to control a vehicle. Outsiders, such as engineers, designers and managers, may be able to help or hinder this goal.

Man as information processor

Because human presence makes systems inherently open-ended, situational uncertainty has been identified as a source of stress. Quantitative measures of information, derived originally by communications engineers, have given good service – in the controllable 'closed-loop' laboratory. Yet the same experiments also showed up the almost infinitely extendable capacity of man to process successfully ever larger 'chunks' of 'bits' (Crossman, 1955). One aspect of the model is that it assumes man to have only a 'limited channel capacity'. To be practically useful, reliable measurement would be essential because interfaces need to be designed so that they deliver just the right amount and range of information as will reduce the operator's uncertainty to levels at which he can cope safely. Unfortunately, real-life cases of intuitively obvious overload, e.g. air traffic controllers, have so far defied accurate measurement.

In fact, the transport operator uses not only whatever information the design engineers care to present to him. Branton (1978) showed in the relatively simple case of the train driver that he greatly depends on extraneous information, both currently perceived and recalled from the past. The real process is genuine guesswork, intelligent anticipation, forecasting of system states, all actions in the future tense. Unless one is able to specify this information, the value of conceptualizing the transport operator as information processor is limited. This does not detract from the virtues of that model; it broke away from behaviourism altogether and highlighted the role of uncertainty in the control of behaviour.

Conceptual limitations to adaptation

The phenomena under observation are boredom, monotony and sensory deprivation (cf. O'Hanlon, 1981). It is noteworthy that all these are closely related to those fluctuations of consciousness or awareness which are manifestations of the general function of adaptation, the familiar and ubiquitous faculty of all sense organs. Since organs do not directly perceive objects, but immediately merely report changes in their surroundings, they 'adapt to' invariance in the rate of such change. From the present viewpoint, therefore, regularity and familiarity of stimulation are the enemies of the senses.

Adaptation implies value judgements

Adaptation is not an all-or-nothing function but varies infinitely in degree; though this is virtually impossible to measure, the slightest change evokes response. However, unless there is a clear response, the determination of a point at which adaptation 'breaks down' must be arbitrary, just as it must be a matter of judgement when a state of arousal turns from 'sufficiently high' into 'harmful stress'. Equally, 'adverse' environmental conditions were said earlier to be compensated by homeostatic mechanisms. Such statements must be recognised for what they are, namely value judgements. There is no need to ban value judgements from discussion. They will anyhow only re-emerge in hidden form. As long as they are acknowledged as such, they help to clarify matters. The danger comes when they, in their turn, are 'adapted to' and so taken for granted that they may be unknowingly misapplied.

Uncertainty and self-generated information

How then does a person in a compulsively restricted situation, say, a driving cab or cockpit, respond to uncertainty? Analysis of field study protocols shows that his 'adaptive response' is to create his own information. More or less con-sciously, he takes the risk that his quasi-information may not conform to the real state of affairs. To cover that gap between his 'imagination' and the future 'real world outside', he continuously seeks for further information from the environ-ment. The term 'adaptation' now changes its meaning from passive compliance to active endeavour. He actively scans the surrounds with all perceptual modes at his disposal and when it is poor, say in fog, he strains his perceptual apparatus to the highest degree. If the environment is impoverished by imposition of artefacts, as in the uniformity of motorways or railway lineside equipment, or even by the well-intended soundproofing of his work station, he takes greater risks and guesses where he does not know for certain.

It may seem rather irrelevant or strange to introduce the distinction between 'imaginary' and 'real' into a discussion of the practicalities of transport operation. But the relevance is justified as it arises particularly from common experience in studies in which mental underload is reported. The phrases repeatedly used are 'mind-wandering', 'day-dreaming', (Oswald; Kripke), 'absent-mindedness' (Reason), 'neglect of surroundings' (Broadbent), 'shift of attention from environment to internal events' 'relaxation of vigilance' and 'under-arousal'. If the senses are lulled, interests and controls are diminished and the ever ongoing mental activity seeks its material from fantasy by a random walk among 'internal representations'. Control over thought processes necessar-ily involves some monitoring of the surroundings. The more intense and exact the control over the exteroceptive senses, the greater the influence of the 'real world' on the person. At certain times of day – and night – a natural struggle occurs between a 'realm of fantasy' and one of 'reality'. In this struggle, all will depend for the operator's reliability on whether mental activity is more or less controlled.

'Mini-panics'

Field observations of repetitive manual work and boring, vigilance-type inactivity have directed this author's attention to the occurrence of spasmodic episodes of hyperactivity, anxiety and stress, tending to arise at intervals of 90–100 minutes. Branton and Oborne (1979) characterize these episodes of day-dreaming followed by sudden actions as 'mini-panics'. Self-reports by articulate professional operators make it clear that they are aware after the event of their minds having wandered from the task. They know that there was a time gap during which their mental activity had slipped out of control but are quite uncertain about the duration of that gap. Depending on the retrievability from any mishaps during that moment, the uncertainty will be more or less stressful. Even though these are experienced operators, they do not seem to be able to anticipate or prevent these lapses.

Ultradian rhythms and postural control

While the stress aspect has been dealt with by these authors, the reliability of alertness needs further discussion. Once it was realized how similar these day-dreaming episodes were to the sleep stage REM cycles of dreaming, the obvious question arose whether such cyclical events did not generally and naturally extend into daytime. In the literature of sleep research and on biorhythms some empirical and clinical evidence was found. It can be subsumed under the heading of 'ultradian rhythms' in physiological changes throughout wakefulness as well as sleep (Oswald et al., 1970; Othmer et al., 1969; Kripke, 1974; Lavie, 1979; and others). The findings are that, in the right circumstances of enforced inactivity throughout the waking day, almost everyone is subject to such changes in arousal at about 90–100 minute intervals. They are reported to be regularly accompanied by brief spells of day dreaming. Of course, body movements are largely suppressed during sleep. But even if one is not asleep, yet in a well-supported posture, whether seated or horizontal, the anti-gravity muscle action control is 'switched off'. If in addition, the eyelids close and visual perception is suppressed, the person easily enters into twilight states in which the balance between attention to the environment and self-awareness becomes unstable. These transitional (hypnagogic and hypnapompic) states are frequently associated with hallucinations and with highly original and creative thoughts, in other words, with fantasies. A connection between secure suspension of the body and reduced information intake on the one hand and enhanced abstract mental activity on the other is suggested. Conversely, it is arguable that the two factors which normally prevent one from being conscious of these states of diminished external awareness, both in daytime and during the transitions to sleep, are the active search for stimulation, particularly visual, and antigravity muscular work. In short, day dreams are less likely to occur if one keeps oneself upright and actively seeks to occupy eyes and ears.

To sum up this section: It was said at the start that in transport, man pits himself against nature, but that this alone is probably not the direct source of stress. By various considerations, these sources have now been narrowed down to the need for knowing what is 'real'. In a boring task situation, the operator should be able to control his mental process so as to be as certain as possible that he can distinguish between what is real and what he imagines at any given moment, so that he can act accordingly. The uncertainty uncovered in this case refers not to the overcoming of (inanimate) natural forces, but to the inevitable interdependence with other persons. Because others rely on the operator's constancy of awareness of them and their interests, the critical factor in this type of stress is social.

An explanatory model

While the foregoing conjectures about natural fluctuations in awareness explain some of the observed behaviour, the self-reports of stress can best be explained by the assumption that operators actually feel responsible for their human charges. They have made it their purpose to conduct themselves safely. A philosophical, meta-psychological case for purposive explanation of a broader spectrum of human behaviour has been set out briefly elsewhere (Branton, 1982). For the present, a subsidiary model is proposed which takes the form of a strictly autonomously controlled person, necessarily possessing value standards and interests in social relations, perpetually seeking and evaluating information from the surroudings. The search varies in intensity, depending on arousal level. This level normally fluctuates in cycles of about 100 minutes. The model concerns the form of mental operations, their contents being material either taken immediately from the surrounding world or from stored, primarily emotive experiences. The explanation is purposive, rather than causal, as it is argued that the thoughts which determine behaviour are forecasts of future states of affairs and their consequences for the person, rather than past experiences in themselves of speculative origin. Needless to say, this statement can only be the briefest of telegram-style sketches in the space available.

Methods appropriate to new parameters

To find ways out of this conceptual maze, one guiding thread may be offered to the researcher. It is to adopt an 'epistemic' strategy: to ask himself in the first place at each stage, 'How do I come to know whether this or that statement by the operator is true?' 'What is the source of my knowledge?' 'How direct or indirect is my perception of the measurement?' 'How far is my conjecture based on analogy and how far does it penetrate to the "real thing" ? Having thus become conscious of their own inevitable bias, observers (and their readers) are better able to speculate on their Subjects' knowledge, values and actual purposes.

Even though it is necessary to trust in self-reports elicited post hoc that will not be enough; the endeavour must be to obtain convergent information from a number of disciplines and angles. The matter under investigation is serious enough to demand in addition a search for the precursors of states of low arousal and of loss of control.

If reliability depends on the operator's control over himself as well as over the system as a whole, parameters will be more informative the more they are sensitive to changes in the control of his behaviour. This is why energy consumption measures were rejected earlier on as unrepresentative of mental work. Control can be exerted either over task related 'cognitive' and potentially cognitive elements of actions, e.g. hand movements, or over behaviour of which the Subject is not normally immediately aware at all. The latter category includes such control functions as affect perception (e.g. eye movements, eye blinks, width of the papebral fissure, increases in latency of overt 'orienting responses') and postural maintenance (e.g. body sway, standing or sitting, and tonic head/neck muscle activity). In consonance with the autonomy view, records of spontaneous behaviour and physiological activity should be particularly informative – provided their meaning and function can be hypothesized. Examples are pupillary movement and limb tremor, but there are likely to be others like transients in EEG, ECG and EMG, which still await meaningful interpretation.

An important implication of the insight into the continuously floating threshold of awareness of the person about to be observed is the requirement that recording and measurement of behaviour and functions should be unobtrusive and non-invasive. Here the ever present capacity to adapt may actually help the researcher to overcome the Subject's self-consciousness. In this author's experience, Subjects quickly adapt to the observer's presence in quite close proximity provided the latter studiously avoids attracting attention. The slightest movement, let alone an overt intrusion into the Subject's world, say, to administer a reaction-time test or a questionnaire, are bound to awaken him. This would crucially defeat the purpose of the whole exercise, namely to capture and record a reality of internal events expressed in unprovoked behaviour.

Other methodological requirements will depend on specific circumstances and the attitudes and co-operation of managements and organisations involved. It can, however, be said in conclusion that, given the premise of this paper – the individual responsibility of the operator – managements and organisations must make an increased effort to allow operators to develop their potential by design for enhancing self-awareness and confidence in their own conscious judgments. Social support, if it is given sincerely by a management committed to safety and reliability, will enhance the confidence of operators in their own conscious judgments. To prove this proposition, a specific effort must be directed towards a fresh attempt at purposive explanation of operator behaviour. The practical usefulness of the attempt will be to enhance understanding of one's own body functions, which is likely to arouse one to act before drowsiness or stress become overwhelming.

Summary of practical recommendations

A) Organisational:

1. Active programme of stress/fatigue prevention, with periodical review built into system.
2. Environmental Design
   (a) avoiding rhythmical 'lulling' stimulation in sight or sound
   (b) introducing mild climatic variations, including frequent air changes and humidity/ionisation balance
3. Positive provision of secondary activities not competing with main task. Operations designed for some walking about and standing as well as sitting
4. Job enrichment, alternation and rotation
5. Re-creation of positive interest in the job
6. Equipment design and standardisation through representative experimental validation.

B) Individual

1. Consciously self-induced changes of posture (abt. one per minute) especially when low on the arousal curve
2. Deep breathing for transport of oxygen to brain
3. Occasional check of time on task – rather then mere clock time – to establish position on arousal curve in relation to 100 minute cycle
4. Intake of glucose/fructose or caffeine
5. Cooling of head, ears, wrists, ankles to counteract too warm or stale environmental conditions.

References

Bainbridge, L., 1978, The Process Controller. In *The Study of Real Skills I: Analysis of practical skills* Singleton, W. T. (Ed), Lancaster: MTP Press.

Branton, P., 1978, The Train Driver. In '*The Study of Real Skills I: Analysis of Practical Skills*' Singleton, W. T. (Ed), Lancaster: MTP Press.

Branton, P., 1982. The Use of Critique in Meta-Psychology. *Ratio*, **24**, 1–13

Branton, P. and Oborne, D., 1979, A behavioural study of anaesthetists at work, in *Psychology and Medicine* Oborne, D., Gruneberg, M. and Eisler, J. R. (Eds), Academic Press, London.

Broadbent, D. E., 1953, Neglect of the Surroundings in Relation to Fatigue Decrements in Output. In *Fatigue: A Symposium*, Floyd, W. R. and Welford, A. T. (Eds), London: H. K. Lewis.

Crossman, E. R. F. W., 1955, The Measurement of Discriminability. *Quart. J. Exp. Psychol.* **7**, 176–195.

Friedman, M. and Rosenman, R. H., 1974, *Type A Behaviour and your Heart*. New York: Knopf.

Kripke, D. F. 1974., Ultradian Rhythms in Sleep and Wakefulness. In *Advances in Sleep Research* Vol.1. Weitzman E. D. (Ed), London: J. Wiley.

Lavie, P., 1979., Ultradian Rhythms in Alertness – A pupillometric study, *Biological Psychology*, **9**, 46–62.

Lefcourt, H. M., 1966, Internal versus External Control of Reinforcement. *Psychol. Bull.*, 65, 206–220.

Nicholson, A. N., Hill, L. E., Borland, R. G. and Ferres, H., 1970, Activity of the Nervous System during the Let-down, Approach and Landing: A Study of Short Duration High Workload. *Aerospace Medicine*, **41**, 436–446.

O'Hanlon, J. F., 1981, Boredom: Practical consequences and a theory, *Acta Psychologica*, **49**, 53–82.

Oswald, I., Merrington, J. and Lewis, H., 1970, Cyclical 'On Demand' Oral Intake by Adults. *Nature*, **225**, 959–960.

Othmer, E., Hayden, M. P. and Segelbaum, R., 1969, Encephalic cycles during sleep and wakefulness in humans, *Science*, **164**, 447–449.

Rotter, J. B., 1976, Generalized expectancies of internal versus external control of reinforcements, *Psychol. Monogr*, **80**, (1. whole number 609).

Sanders, A. F., 1981, Stress and Human Performance: A working model and some applications, in *Machine Pacing and Occupational Stress*, Salvendy G. and Smith M. J. (Eds), London: Taylor & Francis.

Chapter 19

VDU stress: Is 'Houston Man' addicted, bored or a mystic?

P. Branton and P. Shipley

Paper presented to International Scientific Conference on Work with Display Units, Stockholm, 12–15 May 1986

This paper was written just as the American space shuttle, Challenger, blew apart before the eyes of millions of TV viewers. The picture showed the rows of incredulous and helpless 'controllers' at Houston Mission Control Centre. This event reinforced our resolve to investigate more closely the ways in which VDUs form our perceptions of the real world. No doubt, this and similar disasters greatly diminish people's faith in technologists to control and master, not merely nature, but the whole world of artefacts which may run away and destroy their makers. It seems that VDU work creates an entirely unexplored social-psychological environment. If new environments create new forms of life, will future interaction between man and machine create a new species – Houston Man? Perhaps his VDU screen presents an illusory picture of the true condition of any plant or process deemed to be under control? Glued to his screen, how far can he actually control the reality out there?

This VDU obsession exposes something deep-rooted in us: a motive to be in control of events. There are many other VDU 'gamblers', not only those in amusement arcades, and they seem to *play* rather than work at their screen. The TV screen, too, evokes in people all over the world a passive visual fascination, regardless of the content of the message being transmitted (Postman, 1986). Moreover, this addictive effect appears not only in the trivial use of electronics. Just as striking to the unbiased observer is the utter absorption of computer programmers, systems analysts and various other types of designers with screen and keyboard, behaving much like compulsive gamblers in their 'quest for certainty', who get a noradrenalin kick out of hitting the buttons.

Our interest is not so much in what is actually shown *on the screen* but in what is behind it, in what the display is supposed to represent to the operator. The processes to be controlled are at least one step removed from the controller's direct experience. The stress arises when these displays can no longer be trusted. The very remoteness from the end product generates feelings of helplessness, a condition often reported in the literature on stress. Some individuals have the capacity to sense this intuitively before a breakdown ('malfunction') occurs to confirm their suspicions.

The stress is amplified when the controller works in social isolation and is unable to check out these suspicions with others. The unintelligibility of the machine adds to this mistrust. Strong negative feelings get in the way of undivided attention and minds wander off the job. When the mind wanders, one may endanger oneself and/or others, e.g. in a power station control room. Also, the machine's 'time scale' functions differ from those of the controller, who is rarely genuinely self-paced but must adapt to the pace of the machine. Machine pacing, on the other hand, entrains the mind and disrupts natural body and mental rhythms. (True, some people claim to thrive on monotony, but we consider this to be merely an escape from social reality, and escape then becomes the attitude while being at work. Without commitment and positive feelings, attention will again wander.)

Even if the programmer succeeds in embodying his ideas of control in the screen displays, these are not necessarily also in the mind of the actual controller, nor in some kind of 'objective' reality. It represents 'manufactured reality' and is only an analogue and often a crude one. The control is sought over something located in the mental space behind the screen. It is therefore a dangerous illusion to abdicate responsibility to machines and to expect not to be found out. Equally, current talk of 'zero-error probability' is as unrealistic as 'designing the human out of the system'.

Stress pathologies result from this old problem in ergonomics: design of machines is too often based on faulty models of the people operating them. If that model is taken from the purely physical sciences, it is bound to be inadequate. 'Natural' scientists probably have their model of the 'uncertain' world correct as far as its inanimate part is concerned. But a realistic model of man as an autonomous being generating his own certainties in such an uncertain world has yet to be drawn up. Unlike any previous work situation, though, human interaction with electronic work-places sharpens any conflict that may exist between controllers and their artefacts. To place the locus of decision-making at the interface outside the person is incompatible with people's need and capacity for autonomy. In our view, a major source of stress is awareness of such incompatibilities, however vague it may be.

We propose a rational model of man as an inherently well-intentioned, responsible, self-aware and purposive social actor, interacting with other people to share decision-making in a complex and uncertain world, who also interacts with – and is potentially frustrated by – artefacts which are inadequate analogues of fallible systems. Evidence for such a view, related to stress at a variety of workplaces, exists in a wide range of literature on skilled performance in process control and transport (e.g. on aircraft pilotage by Nicholson, on train drivers by Akerstedt and Torsvall, on ships' pilots by Shipley, and others in Wegmann (1984). The practical application of this model of the operator under VDU stress is the purpose of our investigation.

More precisely, our assumptions about controller-operators as purposive persons are that:

(a) they are at their best when interacting with a machine as an extension of themselves (this requires a good new theory of models-of-self);
(b) only if the operator knows his purposes accurately, can he perform the job successfully;
(c) only when we know accurately what success consists of, can we successfully design the job (this follows from control theory paradigm).

Conceptually, the opposite of a state of stress is assumed to be a state of relaxed well-being. In Western society, relaxation has become linked with meditation practices which originated in the East but were not thought of there as remedies for the industrial and managerial kinds of stress experienced here. In outline our approach to the problems raised by this use of meditation is to:

(a) analyse the subjective notions of control over the 'mental space' behind the screen as actually held by operators;
(b) extract from the various methods of meditation those elements which enhance self-awareness and self-control, without divorce from reality (which includes the inevitability of social relations);
(c) explore ways of generating this awareness and personal control over behaviour in relation to stress from monotony and boredom (by entering into the controller's personal aims, values and purposes as expressed in observed behaviour);
(d) comparing them with the systems designers' aims.

Many researchers in the field of work-stress in general and stress arising from mental work with display units in particular, have found the use of meditation to be effective in reducing stress. But this may only be symptomatic treatment. It may not solve deeper problems of stress because it cannot be reconciled with basically mechanistic theories of stress. We would challenge such models and hope to show that the views are, in their own way, just as 'mystical' as certain Eastern practices for varying one's self-consciousness.

Psycho-physiological 'mechanisms' are a myth if by mechanical is meant a non-discretionary automatism: No mechanical levers or cogwheels are f)undcin the head. No absolute law of complete causal determinism applies, no rule of exclusively one-to-one relations between cause and effect, no (hypothetical) imperative that any one specific action must follow from one thought. Instead, the rule is an optative, i.e. always with a possibility of choices of action. For this very reason, the unrestricted use of the Newtonian model of the physical sciences is inappropriate to theories of human action. If not altogether wrong, the model as currently used in the human sciences is certainly applied at a wrong level.

Taking the Swedish pioneer work as a starting point (Johansson *et al.*, 1978), explanations of these data are now sought in a wider context. Evidence, we propose may be uncovered in quite ordinary behaviour (so much part of everyday experience as to be completely overlooked), particularly if systematic but unobtrusive observations can be made by modern techniques in naturalistic settings, or appropriate simulations of them. Recorded events may then be explained in operational terms with current knowledge of physiological, cybernetic and cognitive functions. Observations of common behaviour include: sitting posture during hours of active VDU work, mind-wandering and inability to concentrate while watching for signals on screens and nodding-off in long-distance driving. Field studies of boring situations report evidence. Operational jargon, such as train drivers' fear of 'passing a signal at danger' and air traffic controllers' fear of 'losing the picture', we suggest, are indicators of these (psycho-physiological) stresses. Awareness of responsibility for potential hazard by these operators constitutes an important source of stress experienced by those at risk.

Research being developed in our laboratory complements our earlier field studies of responsible occupational groups. Our reasoning is informed by a psycho-physiological theory combining findings from bio-rhythmic studies with our own work on what we term the 'responsibility factor'. The approach is well documented in a study of hospital anaesthetists (Branton and Oborne, 1979). In summary, the theory presumes that:

(a) autonomous people cope with stress continuously by self-control and self-motivation;
(b) a common type of stress is that imposed by transient incompatibility between the demands of the task and bodily events, e.g. 'nodding-off'; these are often very brief;
(c) compensatory efforts out of a sense of personal responsibility can succeed in overriding such bodily events and needs.

In the case of the anaesthetists, and of train drivers (Akerstedt and Torsvall, 1984), there was clear evidence of 'micro-sleeps' recurring during prolonged monotony. These 'micro-sleeps' recur with a periodicity of roughly 90–100 minutes (following a basic rest activity cycle (BRAC) proposed earlier by Kleitman). These were sensed by the anaesthetists via unconscious cues from postural feed-back mechanisms – on return to full consciousness after nodding-off. The results were typically 'mini-panics' as the conscientious dozers were jerked back into action when they realised that the patient might have been deprived of oxygen during the moment of inattention. It is possible to see this phenomenon on speeded-up video recordings. We propose that these risks are much greater in monotonous conditions and in sleep-deprived operators, whether they are in control rooms or on the move.

Other researchers have documented fluctuations in concentration at various work stations. We believe that more substantial evidence for the existence of ultradian rhythms, having the capacity to affect attentional qualities, has been less forthcoming because these rhythms simply have not been looked for; procedures for EEG analysis for example are not normally suited for such 'fine-grain' studies.

Findings about VDU stress have often been confused because of basic differences in mental operations required by VDU tasks. VDUs are being used in workplaces in two fundamentally different ways which seem to have different implications for health, well-being, safety and performance. The conversational or dialogue mode deploys the VDU as a creative aid, literally a tool, for the user as innovator, decision-maker and controller. We have tried to describe the dangerous addictive and illusory risks which sometimes may be associated with VDU use as in our reference above to 'Houston Man'. It is, however, a vastly contrasting application, the repetitive data input or data entry mode, increasingly widespread in industry and commerce, which currently raises the majority of problems discussed in the VDU stress literature and which is the basis of our present experimentation. This is the Tayloristic conveyor-belt in a new guise and it presents work load and monotony problems of its own. They are the 'new technology sweatshops' of the latest wave of industrialisation.

In most cases speed and variety of electronic information display creates a sharp divide into those who control and those who are controlled, those who have exciting jobs and those whose tasks are boring. Controllers and decision-makers are supposedly aware of the complex consequences of their every action, whilst their 'slaves in the network' have strictly limited discretion. The former are said to suffer from mental overload, the latter from 'underload'. Yet both can experience sufficient stress to incapacitate their performance and perhaps affect their wellbeing. Limiting exposure by the manipulation of rest periods is a common intervention strategy, but that does not get at the root source of the hazard and perpetuates badly-designed jobs.

Our experimentation is similar to that described at this conference by Floru *et al.* (1985). To test our theory we have developed a simulated monotonous low-discretion job at the VDU and the operator is attached to a portable EEG monitor which continuously displays the various frequencies of currents for each hemisphere separately. The operator keys-in the numbers from simulated bank cheques for a prolonged time in a socially-isolated and monotonous environment, constantly monitored by video cameras focused on both the operator and the EEG display. Operators are psychologically tested beforehand, are working an 'evening shift', and complete a stress-arousal 'mood adjective check list' before and after. Performance at this task will be contrasted later with a VDU game to represent a non-boring, non-monotonous task.

Acknowledgement

Thanks are due to Mr George Faria for setting up this study and Mr Matthew Balantine for writing the Performance Measurement Programme.

References

Akerstedt, T. and Torsvall, L., 1984, Continuous electrophysical recording, p. 567–583, in Wegmann, H. M. (ed.) *Breakdown in Human Adaptation to Stress*, Vol. 1, Part 2, Dordrecht, Netherlands: Nijhoff, for Commission of the European Communities.

Branton, P. and Oborne, D. J., 1979, A behavioural study of anaesthetists at work, in Oborne, D., Gruneberg, M. and Eiser, J. R. (eds.) *Psychology and Medicine*, London: Academic Press.

Floru, R., Cail, F., and Elias, R., 1985, Psychophysiological changes during a VDU repetitive task, *Ergonomics*, **28**, 1455–1468.

Johansson, G., Aronsson, G. and Lindström, B. O., 1978, Social Psychological and Neuroendocrine Reactions in Highly Mechanized work, *Ergonomics*, **21**, p. 583–599.

Postman, N., 1986, *Amusing Ourselves to Death*, London: Heinemann.

Wegmann, H. M. (ed.), 1984, *Breakdown in Human Adaptation to Stress*, Vol. 1, Part 2, Dordrecht, Netherlands: Nijhoff, for Commission of the European Communities.

Chapter 20

In praise of ergonomics – a personal perspective

Paul Branton

*International Reviews of Ergonomics, 1987, **1**, 1–20*

Abstract: This paper describes, in a personal way, a search for the links which could possibly bind together the many, apparently quite disparate areas, to which Ergonomists devote their efforts. Naturalistic observation of people's behaviour highlights some puzzles that cry out for explanation. There is, for instance, their inability to verbalize even their own most simple skilled actions. To perform as they do, their world must have been represented internally in certain forms. Yet to be explained by neuropsychology, that world must be replicated in all its complexity by continuous processes in neural (brain) tissue, and the mental operations upon these representations must be both extremely precise and yet flexible enough to foresee future states of man–machine interactive system. It is shown that valid explanations must go beyond current preoccupations with computer analogies. The explanatory power of the concepts of autonomy and purposivity are explored. A broad concern with principles is necessary if responsible Ergonomists are to provide help in overcoming the hazards of Chernobyl and the Space Shuttle. Ergonomics, with its emphasis on practical application to normal humans, is seen as the life science *par excellence*.

Introduction

The lead-in to an annual journal reviewing current trends in research and practice demands a special panoramic presentation, a blend of the past with glances into the future. This synopsis of an Ergonomist's occupation is necessarily personal and impressionistic, but no less scientific for that, I hope. Rather than strictly chronological, I have structured this account thematically, used insights gained from various assignments that came my way, drawn out puzzling and unresolved experiences, and tried to integrate it all into a consistent, operational explanation of humans at work. Readers are warned that this is no light bed-time reading. I have deliberately kept discussion at a conceptual level, so that others may build on this towards new breakthroughs when their chance comes along.

My first praise of ergonomics is for the breadth of its possibilities; it constantly challenges me *to consider the whole person*. Whenever I apply ergonomics to real life, every specific reductive solution to a problem soon raises another in an adjoining field of the human sciences. My second praise is that it starts with *normality* and works towards the limits of human capacity, rather than starting from the abnormal and unusual. This was a good corrective to a long held impression, dating from my youth in Vienna after listening to some of Freud's lectures, that 'it's all in the mind' and that psycho-analysis explained most of psychology and physiology. My third praise is for the practical significance of ergonomics. By practical, I mean something which helps to answer questions of what should be done. The lay person, told that ergonomists study how to avoid 'human errors', immediately understands the social useful-ness. Other disciplines may serve as technologies only, i.e., as mere adjuncts to theoretical science. If anything, Ergonomics as a technology is explicitly in the business of fitting technology to its users. Its objective, its purpose, is man. As a research model in its own right, Ergonomics deals with hypothetical imperatives of the 'if . . . then' kind, yet its practitioners can never altogether escape from facing questions like, 'What is it all for?' 'What are the values of the system?' More praises will be sung presently, but not idly or without intent. Ideally, some *Principles of Ergonomics* might crystallize themselves.

It was by sheer luck that I came to ergonomics early in my academic career. Previous experience in industry and in the war had turned me to practical matters, yet I always had a speculative streak in me. Ergonomics, with its emphasis on concrete application, virtually forced me to compare theory with reality and adopt an entirely fresh approach to puzzling facts, revealed by actually observing people's behaviour. None of the then current theories offered adequate explanations and so the new angle of view led me almost inevitably to attempt constructing my own abstract frame of reference. When applied consistently, it became a coherent account of purposivity in the actions of men and women, especially when at work. By purposivity I mean that our actions are determined by the future, rather than by the past. Many things, like skilled behaviour, are not merely 'goal directed' but are best explained by asserting that we possess a specific capacity or faculty to represent to ourselves internally our own future actions – and their consequences – *before* we actually carry them out. It is one of those puzzles that we manipulate events that have not yet actually happened; and we seem to do it quite successfully. That this view of purposivity has considerable explanatory power will be shown later.

Sitting in reality

I was driven to this explanation soon after I began to study sitting and seats – the 'folklore' of ergonomics. The old standby, anthropometric dimensions, was clearly insufficient to predict whether a chair would have that elusive quality, 'comfort'. This much had just then been shown in a monumental study in which, among others, a dozen reputed 'ergonomics experts' had sat on (and thereby 'tasted') a set of specific chairs conforming to British Standard dimensions (Shackel *et al.*, 1969). To everybody's dismay not one of the experts was able to predict which chair would be subjectively preferred by a larger group of lay persons.

So I searched for objective measurables, other than anthropometrics, by extensive observations of unselfconscious sitters in large numbers (e.g., Branton and Grayson, 1967). The ever recurring questions were, 'What are they *really up to*?' 'How many different postures did they adopt, for how long, and why the change?' We found that different seats tended to induce certain postures more often than others; seats as well as postures could then be investigated for their mechanical and physiological characteristics. One of the most intriguing questions was 'Why do people sit with crossed knees or legs?' To which the short answer turned out to be that this manoeuvre imparts considerably greater stability to pelvis and trunk. The most probable reason for adopting and maintaining certain sitting postures is to optimize an overall balance between stability and mobility appropriate to the purpose one is pursuing at the time. Normally, purpose determines postural behaviour. Thus, any observer who watches closely large numbers of sitters will almost always find that the bearing of the head and the line of gaze determine the rest of the bodily disposition in the upright.

Of course, these generalizations are the accumulated results of years of research and practice (Branton, 1969a), during which strange things were experienced and much fun was had. We encountered some facts as strange as the science fiction in which people perform experiments upon themselves. For instance, in an unguarded moment, the early pioneer of seat research, Bengt Akerblom, revealed in conversation that he himself was the subject of an experiment in which two 23 cm long steel pins were driven into a man's ilium and great trochanter to obtain X-ray pictures so as to measure most accurately the degree of pelvic tilt during sitting down (Akerblom, 1948, Fig. 33). The least I could do to emulate such courage was to sit for some years at my desk on a self-designed contraption, a 'minimal sitting-body suspension system'. It consisted of two padded boards (about 20 cm × 10 cm), one horizontally fixed, to support the ischial tuberosities, the other vertical, about 15 cm above and behind this for the top of the superior posterior iliac crest to lean against (see Figure 20.1). That seat was perfectly comfortable and I have not suffered from spinal trouble. (By the way, these are the really critical anatomical points of essential support, the dimensions of which can readily be built into any seat.)

Many practical and theoretical points in seat design arose from these relatively mundane observations. But above all, I became initiated into the mysteries of maintaining the upright posture. The search for rational explanations of such mysteries became the springboard for interests which ranged well beyond what one would expect from work on chair comfort. It took me to research into accidents, into stress due to boredom, and into fatigue and sleep. The connection is this: everyone knows that people do not usually sleep standing up or sitting upright, and that they 'nod off' when seated comfortably. The mystery was that subjects tend to relax more easily (at certain times of the day) the further the head is behind a vertical point above the hip joint – provided one

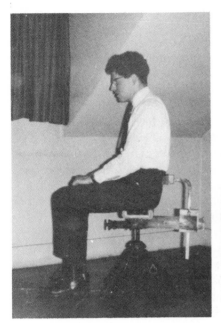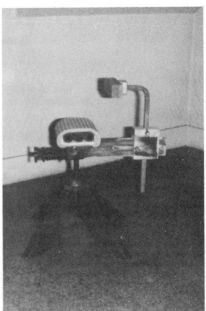

Figure 20.1 The author's minimal sitting-body suspension system

is securely supported. Backrest angles of 30° or more from the vertical seem to be 'drowse-inducing'. If reduced arousal and sleep are inversely related to the angle of posture, what is it about uprightness that keeps one aroused? Conversely, what is absent in sleep that is present in the waking state? The answer to the second question is far from trivial and its full import took a long time to become clear. It is, of course, the continuous fight against gravity which pervades the whole body and its psycho-physiological control system. I had tried to show just how unstable the sitting body is (Branton, 1969b). Could it be that the undoubted dynamic instability of the body in standing and sitting directly affects alertness and consciousness?

Ergonomists, more than most other biologically oriented scientists, face the baffling fact that people, say in power station control rooms, are undeniably tired after a day's sitting down. The physical energy used while sitting, as measured with a Douglas Bag, is negligible. Therefore, the undoubted tiredness after a hard day on the office chair must be the result of something other than expenditure of physical energy. Whatever that may be, it is also well outside normal awareness.

Here, ergonomics spills over into cybernetics, and we know from control theory that feedback loops can control a large mass without expending much physical energy. Could it be that keeping upright costs more in terms of control effort than in expenditure of physical energy? Is it this which makes one feel tired? The distinction between two kinds of energy may help to explain the nature of 'mental fatigue' in other contexts. Thus I was gradually led to study the fatigue aspects in the work of anaesthetists (Branton and Oborne, 1979) and the boredom of VDT operators and data input clerks (Branton and Shipley, 1986).

Beyond direct observation

Quite apart from the anatomical, physiological and bio-dynamical complexities, I personally find two aspects of sitting and standing postures most remarkable: they are mostly adopted unselfconsciously and, although they are plain for all to see, the reasons, or rather purposes, for which they are taken up are not at all obvious. Whatever social meaning various postures may be thought to have is not the point here.

Postures are rarely contrived, hardly ever deliberately taken up and wholly beyond one's awareness, except when one's attention is drawn to them after the event. Nobody says, 'Now I am going to cross my ankles', or, 'Now I will put my hand on the armrest.' (Yet no one outside the person can tell whether a posture was taken up deliberately or not.) The records definitely point to the presence of processes, not directly observable, perhaps not observable at all by instruments, but which must necessarily be going on out of sight *to produce that which is indubitably observed.* The records represent the circumstantial evidence.

To find out more about the reasons for what goes on inside during such ordinary activities as sitting, description and proposed explanation must be in operational terms, rather than by developmental or evolutionary theory. Here we want to know how things work, not where they come from. And we want to understand these operations from more than one angle or level.

An instance of the difficulty of explaining things exclusively at the physical level arose at one time: by use of a modified force platform I hoped to prove that dynamic bodily instability during sitting actually existed and was due to the antagonistic action of skeletal muscles. Indeed, I measured the finest movements of the body, including heartbeat and breathing and found, superimposed, tremor-like movements of the centre of gravity of about 8 Hz. But when I found that there were too many intervening variables to come to a clear-cut conclusion, someone suggested a drastic experiment: to seat a person on a chair on the platform and administer a drug relaxing only the skeletal muscles while keeping the subject awake, to see how the oscillations of the centre of gravity would diminish in that state of relaxation. Comparing the two sets of measurements would at least have demonstrated the contribution of overall muscular activity to the act of simply sitting upright. But the experiment was considered unethical since it was at that time dangerous to administer those drugs. However, it was pointed out to me that the answer was 'intuitively obvious': an 'unconscious' person would just slump in the seat and could not maintain himself upright.

Unfortunately, this answer does not involve actual physical measurements and so did not constitute what was then considered adequate 'scientific proof'. Nowadays, since Kuhn (1970) and Popper (1983), the views of what is scientific have changed and it is generally accepted that every researcher makes some unprovable assertions. Ergonomists are no exception. To deal with inner events which are not directly observable, scientific methods from other disciplines may have to converge and the gaps between them be bridged by imaginative and speculative leaps. So long as it is done explicitly, this can do little harm but may do some good by leading to discoveries and so enlarging the field of research.

> . . . Scientific explanation, whenever it is a discovery, *will* be *the explanation of the known by the unknown.* (Popper, 1983) (Popper's italics.)

I spoke above of 'undoubted dynamic instability of the standing or sitting body'. My conjecture about defence against inherent instability as the purpose of certain postures is a case in point. It may not be testable at the present time, but knowledge of biomechanics and neurophysiology make it very plausible.

From postures to purposes

In this context, the word 'purpose' may sound strange; 'reason', 'goal' or 'aim' might be equally suitable. But these latter words mean something thought out beforehand and that conflicts with what we have found above, namely unselfconsciousness. In this area, choice of words is clearly critical.

These considerations lead me to another virtuous aspect of the breadth of Ergonomics: self-understood acceptance of multi-disciplinary connections, all within the frame of practical human sciences. A discipline which combines an interest in the environment with a search for inner consistency demands a holistic, rather than a limited, answer. If everyday life is all of one piece for a person, however chaotic it may seem, the Ergonomist's descriptions and explanations must also be ever more inclusive and interactionist rather than reductionist (Popper and Eccles, 1977). One of the most sympathetic, yet strictly scientific, cases for a neuropsychological basis of purposivity is found in Granit (1977).

Our finding on postures showed that purposes are not only what we pursue consciously; they are like icebergs, having very large submerged parts which we fail to notice because we take too much for granted. Our conscious purpose might be to look up a word in the dictionary in the next room, get up, go there, find the book, leaf through it, hold it at just the right distance from our eyes until we find the target word; all this belongs to the purposive act, yet is performed with those part-decisions hardly ever deliberated. I shall draw on work done with train drivers to show just how much 'non-conceptual mental work' must be involved in skilful pursuit of purpose. Consciousness is something which some of our colleagues in that most recent branch, cognitive ergonomics, seem to bypass (Bainbridge, 1985). They concern themselves mainly with human functions that are verbalizable and computational, overlooking vast areas of non-linguistic non-conceptual mental activity. Perhaps they hope the problem will solve itself later. However, as I pointed out elsewhere (Branton, 1982), in the process large sectors of real life are ignored, while the problems of explaining such basic functions as selective attention of directed awareness still remain with us. The largest part of our knowledge is still 'tacit knowledge' (Polanyi, 1958), a concept which has an important role in the general case for the purposive view.

Train drivers as forecasters

The idea that purposivity, as internal manipulation of the (near) future, explains observed phenomena better than the usual models which regress to causative factors of the past, was forcibly brought home to me when I investigated on the railways how train drivers controlled their engines (Branton, 1978). On the face of it, that seemed a task relatively easy to describe because it is along the rails and 'one-dimensional', the operator having only one degree of freedom to move, namely accelerate or slow down, to arrive safely and on time. No steering was needed, like in a truck, and certainly no turning and banking, no take-off or landing, like in an aircraft. But quite unlike in other fast moving vehicles, the effect of the train driver's action is much more delayed so that influence over the train's behaviour is so much slower and the time factor is much more critical. The stopping distance of a train moving at 70 mph (110 kph) is about a mile (1.6 km) and the time is often more than a minute, certainly longer and further than can be seen ahead of the cab. The man has to act long before he can see the signal. 'Visual reaction time' is therefore less of a problem than accurate route memory and experience in train handling.

What was officially required was strict compliance with stringent safety rules and accurate timekeeping, besides some technical knowledge. The immense amount of informal and intangible qualities that made up a really good driver did not become clear to me until I had ridden on the footplate for seven years over thousands of miles. By that time, I had begun to understand the 'hidden factor' behind such simple statements as, 'You must stop, not at, but *before* the buffers at Euston'. Or 'You must not pass a signal at danger', i.e., when it is red. The rewards for digging for the principles of this skill were that they revealed themselves to be just as applicable to other vehicle operators as well as to many types of process controllers.

Inability to verbalize

First, a colleague, who was a senior railway operating officer in charge of all the locomotives, and I interviewed hundreds of drivers and inspectors about 'what they really did while driving'. They eloquently describe the route in terms of personal and unofficial cues, rather than in signals and other line-side equipment provided. But by far the most striking thing was that these men, some with 30–40 years experience and with high IQs, were quite unable to put into words how they drove and what they really did, and when they decided to do what they did. To get to the bottom of this puzzle, I embarked on a study by unobtrusive, 'eyeball' observation of their behaviour. I watched what controls they operated, when they did it; I followed as best I could their line of gaze. As was by then second nature to me, I watched their postures, how they moved their feet, operated the 'dead man's pedal' and so on.

This is how I found what I think is a feedback signal from strong brake applications: a slight forward move of the trunk away from the backrest. This was important, because of the slow propagation of the effect of the braking engine on long chains of old coal wagons, in particular on the steep Welsh valley routes. I sometimes measured that interval between action and response to be in the order of half a minute and more. Needless to say, stresses like having to wait so long before he knows whether the brakes have bitten, make the driver's job much more arduous than laymen – or even management – are aware of.

Task variables and their predictability

From the Ergonomist's viewpoint, the next step was to describe the variables in the task in ergonomics terms in such detail that it can be quantified later (see Branton, 1978, in particular tables 8.1 and 8.2). These data formed the basis for defining the conditions under which the driver operated and which provided the background for the next phase, skill analysis. The relative importance of these variables lies in the extent to which they are predictable and whether the exercise of skill depends on immediate perception, on fine sensory-motor control of brake and power handles or on geographical memory. For each of these categories, different kinds of skill are required. Ergonomists know a great deal about the first two and much experimental work has been done under the heading of 'tracking', but the task of applying route knowledge seemed to me relatively unexplored. Bearing in mind the need for devising training instructions, we must know what exactly the operator was expected to recall from his 'working storage'.

Mental operations with trajectories

Of all the considerations involved in describing this clearly complex skill, the most far-reaching one turned on the nature of internal representations and mental operations. When a highly responsible operator performs a task so accurately and safely over time and distance, and mostly without direct or immediate sensory feedback, some very essential data must be represented internally. Simple visual memories in the form of static targets as seen in the past are totally insufficient to explain the accuracy with which drivers bring the trains to a halt just before signals and a yard or two before the buffers at the terminus.

To illustrate the problem of what would need to be represented internally, Figure 20.2 is a schema for the decision to apply the brake on the approach to a signal at red.

> . . . For the driver, point A has no fixed location on the ground. To him, A is variable because he has not only to consider the relatively constant brake

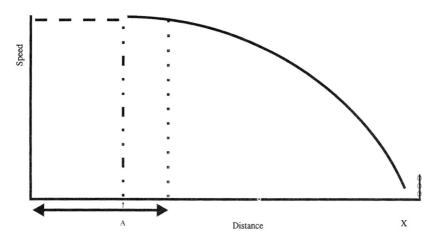

Figure 20.2 A driver's goal-directed braking curve

characteristics at given speeds, but also many other variables; such as train composition and mass, the weather, etc., and even whether he is early or late. [He] must therefore hold in memory exact representations of these variables, as well as of all possible locations for the start of [successful] brake applications (Branton, 1979).

The driver matches his representations of the appropriate variables with a location on the track in the zone around point A and makes his decision well before reaching the zone limit. It is entirely up to him when and where along the track he applies the brake – as long as he stops at the signal when necessary. Because of the accuracy of actual execution, I conjectured then that the mental operations upon the representations are quasi-mathematical, implicit solutions of time/distance equations of trajectories. Of course, these are tacit knowledge, implicit calculations carried out quite outside awareness. Since they are directed at a goal not yet reached, they are essentially purposive.

Purposes represent the future

Like a signal round the next curve in the track, a purpose is always well ahead, not only in distance but, more importantly, in time. At first I thought that since this purpose is almost never within sight the representation must be written, as it were, in the future tense, an intention. Then I realised that such representations would be too rigid to yield the observable successes. Because no two occasions would ever be quite alike, any change in all the many task variables would have to be taken into account before execution. Moreover, there is usually a variety of ways in which to achieve the purpose, and the choice among them must also come before action. Furthermore, the consequences of specific courses of action must also be compared and contribute to the final decision to act. The tense of the foresight would therefore need to be a-little-after-the-future, or as the grammatical phrase has it, *paulo post futurum*. There is nothing strange in an intentional representation; when he learns the route, the man says, 'I must remember this point for the future'. Indeed, I consider that most of our representations are somehow laid down 'with a view to the future'. Learning and memory processes are not just for nostalgia but for forward projection.

Internal representations and mental operations

That such internal representations and processes exist is not in doubt. Where could they be? Surely not at one single point in the head? In what form could they be held? 'Engrams' and structural 'cell assemblies' would be far too rigid. Even the word 'exist' raises philosophical problems. If I had sat in an ivory tower, I might have said it's not my business to philosophize about the existence of thoughts. Instead, it was my business as an Ergonomist to help to improve operating safety and once the questions were posed I could not let go. I did argue with philosophers of mind about Monism and Dualism and how far a physical/materialistic or a psychological/animistic explanation was valid. I toyed with interactionism, but found it did not really account for the inevitable selection of one representation over another. Also, it had not enough room for a scientific approach to values and purposive human behaviour. These latter must naturally also be represented somewhere in the head. How else would a driver know which of his actions was important? The question then turned into, 'What exactly is it that is represented inside the driver?' Some call it memory, others experience, and pass on. I had to conclude that, to act as he did, he must *possess knowledge* of the whole of his 'railway world' in all its complexity, together with the immediate consequences of his own actions, as well as all the postures and movements required for success. He just could not have achieved his skilled action without this 'intelligent knowledge base'. (As will be discussed below, it is possible that an object may not be stored in the head as such but only as related to its features.)

How far is the idea that everybody carries their own universe with them generally valid? If it is so, the anticipatory aspects of all skills could be explained in terms of operations with these indwelling representations. This may be what a respected Ergonomist, L. R. Young (1969), once somewhat mystically called 'precognitions'. I am convinced that the importance of the function of representations is far wider than is usually realised. The mere fact of our having in our head a kind of ready-to-use replica of the outside world makes us relatively independent of our immediate perceptions of the environment and thereby autonomous. By this means we can break the natural chain of causal determination and can become self-determinate. We are then no longer merely reactive, but self-activated. We don't have to do a thing but can 'think it over' before we move. It all may look rather simple to the point of triviality, yet it is fraught with implications for learning theories and training methods. My excuse for this philosophical digression is that it strengthens my confidence in a theoretical framework which may eventually take ergonomics – and other human sciences – into new fields. It may even free one from barren forms of empiricism, to look beyond one's nose. Who knows, some Ergonomists might even be tempted to speculate or philosophize mildly for themselves rather than meekly accept Descartes and Hume. Note that I stress the 'mildly'; the exercise must not be carried to extremes.

At a later stage in my career, when I had a little more time to think, I realised that neither the monist/dualist nor the materialist/mentalist controversies are directly relevant to the present case. I turned to another border-line discipline familiar to Ergonomists, to neuropsychology, for an answer to the question about form and place of representations. My present conjecture as to the form in which these representations are actually held in storage is based mainly on neuropsychological evidence (Pribram, 1971). Considering that, in the driver's case, any achievement of purpose requires muscle action, the most parsimonious form of representation would be in loosely ordered posture and other movement sequences, in what William James called 'enactive' form. It means that no transformation into symbolic or conceptual representations need take place between perceptual and motor processes and that expression may be immediately muscular. For one thing, this would explain why the drivers could not verbalize what they were doing. Also, it is a common observation that very articulate people, even eminent scientists, sometimes underline their speech with gestures, 'thinking with their hands', apparently without being aware that they do this. Being now on the look-out for the unknown, we get here again a glimpse of submerged parts of the iceberg, representations that were initially unconscious but become public by behaviour.

Success must be known before errors can be defined

In the branch known as 'error ergonomics' (cf. Singleton, 1972), much is made of the use of feedback loops, often in exact computational terms. However, in the railway case, the driver is himself an open-loop, nonlinear system, and also operates open-loop. He controls his train only intermittently, rather than continuously, the whole being in effect a ballistic missile proceeding, so-to-speak, from one target signal to the next. Whatever feedback he has must be complex and depend on his watch plus his knowledge of the route. His ability to correct errors and make up for lost time is very limited. An error committed now would not be known for some time ahead. Because of this, to me at least, the compelling experience in the driving cab was one of helplessness to influence one's fate.

An important logistical point, often overlooked, is how arbitrary it can be to designate something as an error. Before errors can at all be corrected, they must first be accurately detected, which can only be done by comparing output via a loop with the original standard input. Only when the input is known very precisely can error be accurately defined and feedback used to achieve success. The point is that what constitutes success must be known first; only then can error be at all determined. Only when one has acted purposively and set a particular goal, can one speak of success or failure and hence of error. This link cannot be stressed enough. Its implications for all human factors work, including computing systems and programming, are profound. For example, I find the specification of what constitutes an error a powerful criterion in assessing the validity of safety systems of all kinds. Knowing that there is virtually no system totally without any human in it, and knowing that 'absolute success' is an indefinable technological fiction, it would be very arrogant to claim for a system that it is error-free. Even the best 'built-in automatic error correction' is useless if it does not provide for 'human error' and specify exactly how it is to be corrected successfully. As Bainbridge (1983) cogently pointed out, 'the more advanced a control system is, . . . the more crucial may be the contribution of the human operator'. The same applies to the designers, supervisors, and maintainers of any man-made systems.

Representations of intended success

Success in skill may be the result of tacit knowledge, but once a purpose has been achieved, it provides hard and indubitable evidence. Success tells us more about what must have been represented internally than any causal analysis of failure. Only when the operator of a system gets his intentional representations right can the man-machine relationship succeed. However, to find out enough about the inner replica of a person's world is very difficult indeed. Maybe we have so far not looked in the right places. As to where to begin looking, I have tried to argue elsewhere about brain locations (Branton, 1985), that objects are 'decomposed' into their features by the brain and stored in dispersed form as attributes. Thus the roundness of an object is detached and may be stored, say, near to representations of circularity and ellipticality. Similarly, that same object's redness may be represented in a cerebral column among its spectral neighbours. Any one particular attribute may be shared with many otherwise disparate objects. Any one object need then not be stored as such but can be cognized by all or some of its attributes. The crucial importance lies with the connections or relations between features because objects then exist in the head, not punctate, but relational to features: a relational system. (How is that for a new approach to 'parallel processing'?)

Such a hypothetical view of a form of storage is therefore much more parsimonious than past theories proposed and it is, in fact, in close accord with recent findings in brain research (cf. Hubel and Wiesel, 1962; Blakemore, 1976). In ergonomic practice, this insight may help to determine experimentally, for instance, the relative merits of colour-coded and differentially shaped control knobs and levers.

The Ergonomist as researcher must have empathy

How difficult, but also how desperately necessary it is to throw light on those inner representations which guide a person's actions became apparent to me as I got involved in the study of stress, in particular under conditions of boredom. As an observer, one is greatly helped then if one has a capacity for empathy with the operator. That does not mean that one should be biased or otherwise favour the particular person observed, but that one should make a deliberate effort to enter into the situation through the eyes and mind of the observed. Since the much vaunted idea of complete objectivity is found wanting in the physical sciences and thereby also in the human ones, only a cautious impartiality can be advocated as antidote to complete subjectivity, emotionalism or relativism. The method of naturalistic, unobtrusive observation and measurement in the field, when practised in Ergonomics, can help considerably to diagnose problems and its heuristic value can reward long, dull hours of close watch.

Stressed by mental overload

Ergonomics has always prided itself that it can measure work load and its limits. For physical effort this may be the case; I found that the problems really started when the non-physical phenomenon of stress had to be assessed in everyday industrial practice. In particular, the whole conceptualization of what constitutes stress seems to stand in the way of breaking new ground, because it is based on a pre-historic model of humans, rather than on present realities. After a brief critique of the model, I shall explore stress in relation to three situations: overload of mental work, underload through boredom and the effects of responsibility.

Traditionally, Cannon's 'fight and flight' syndrome has been used to prove by physiological and biochemical measures that a very high mental workload is stressful. My experience stems from the 1960s, when many railway medical authorities all over the world tested for catecholamines and frightened everybody who got to know the results. It became clear that these measures were not 'fine-grained' enough to allow detailed recommendations to be made in specific cases. They may be all very well as demonstrations that stress actually exists, but the step linking high heart rate and adrenalin/noradrenalin excretion with the claim that all adrenalin and all stress is *harmful* is unjustified.

My objection is two-fold: that the connection is made too uncritically and that a 'normative' concept is tacitly introduced. First, the link is based on intuitive, speculative assumptions of evolutionary history, namely that today's physiological functions are the remnants of prehistoric memories, 'instincts'. Of all the people I had to measure in my time – train drivers, factory workers, airline pilots, managers, anaesthetists – none ever fought anybody in caveman style. Neither had they the slightest chance to use their muscles for fleeing from their work. Second, the connection contains an implicit value judgement. The pejorative implication in stress is tacit. When stress is a bad thing from the outset and it is made so without specifying some limit, the purpose of research is defeated. Now, I have nothing against value judgements – I make them all the time – but I object to their tacit and unacknowledged infiltration. This syndrome has become an outdated paradigm which stops us from thinking further. High time it had a decent burial.

Why then pretend and continue the ritual measurement? In the final analysis I think it is due to the 'dogma of empiricism' (Nelson, 1949), which denies the existence of *synthetic a priori* propositions: tacit, not dependent on sensory experience, somewhat imaginary and unprovable. Yet, the idea of causal determination is just such a proposition, constantly used in human factors research. In this case, we say, 'the cause of stress is . . .' It is tacit because we take it all the time as read that everything has a cause; one rarely asks whether it determines events unequivocally. (Although in accident research, the idea of multiple causation is well known.) It is independent of sense experience (though not necessarily before it in time) as nobody has yet observed a cause determining anything. The idea is synthetic as it brings together two concepts and 'imagines' the necessity of the connection between one cause and one effect. Adrenalin may cause stress but, as Selye frequently pointed out, there is such a thing as 'eustress', a condition desirable for certain purposes. The determination as to what kind of stress is present or desirable depends, as we have seen above, to some extent upon which of his internal, intentional representations the self-determined person chooses and how realistic they are. That this last proposition is ultimately not amenable to positive 'proof' does not cause me many sleepless nights. I have no faith in logical positivism. I have a certain confidence in the truth that I am responsible for what I do and that I could resolve doubts or conflicts by reasoning with other people who may be affected by my actions.

Overload and responsibility

Unfortunately, individuals work within systems in which technology demands more than is within their natural capabilities, over which they, anyhow, have less than full control. The train driver, unlike a car driver on a minor road, cannot suddenly decide to stop and take a rest when he notices that he is tired. He is responsible for the train and may overrule his 'automatic' bodily functions up to a point. Most often, he does not even recognize the point at which he becomes overloaded until too late. Can he be blamed if the system is taking him beyond his limits? If anyone could help to answer this question, it would be an Ergonomist – if only he knew enough. To distinguish between 'beneficial' and 'harmful' stress, it may, for instance, be interesting to know the effects an operator's responsibility has on his own body.

The concept of the responsible person, as applied to transport operators, has been elaborated (Branton, 1984) and has clarified my ideas about ways and means of actively reducing stress. In the literature, neuropsychological evidence for the validity of the concept of responsibility can be found in unexpected places. A good example is in Nicholson *et al.* (1970). Heart rate and finger tremor of a regular airline pilot and his co-pilot were simultaneously recorded during 34 landings. During the let-down, up to about 1000 ft (300 m) above ground, one pilot handed the controls over to the other for the last minute to touch-down. On one occasion, the former pilot's HR was reduced from 85 to 70 within 30 s, whereas the latter's HR rose over the same period from 90 to 125 and then remained for another 40 s at a peak of 136 (only approximate figures in beats per minute are given here). In another case of 'shared workload' at a difficult approach, the first pilot's HR fell from 140 to 96 on handing over the controls, while his partner's HR rose from 80 to 130. (Note that 130 is not unbearably high.) The authors conclude that the increased HR 'in the absence of untoward events' indicates the mental process of computing the landing configuration. However else these data may be interpreted, to me they show the immediate and real effects of what would otherwise be a vague concept, responsibility. Both pilots – and anyone else on the aircraft – are exposed to the same risk of error but only the one in control has the unseen effects. Of course, I cannot prove the connection directly, but equally I cannot imagine a better reason for these physiological changes. My protocols from interviews with train drivers make frequent references to being 'dining-car conscious' or knowing 'instinctively' who they carried behind them. That is why I think that stress contains a large social element.

Stress through boredom

Boring work is another case where the system takes people to their limits. The usual point made in industry is that women are better at such tasks. As if they had no limits! It is, however, unreasonable to put it down to personality differences and to expect that personnel selection will find the right persons, men or women. Aware of the need to specify human limits, Ergonomists must start by assuming that every operator as a person is responsible for his or her own conduct. But the search for human limits arises particularly in situations where the operator is predictably subject to boredom, such as in power station control rooms, where unremitting attention is critical. Usually in these the operators monitor automatic devices in monotonous environments over long hours for rare events like malfunctions and so became naturally under-aroused, inattentive and drowsy. When the rare emergency does arise, they are then less able to cope. Since Chernobyl, all the world knows what we understood for some time: that no system is totally infallible; what is virtually impossible to predict is 'mean time between human failures' (a reliability engineering concept).

Drowsiness as lack of intensity of purpose

The idea of purposivity may help here to find new insights in a renewed attack on this urgent problem. Purposive action is distinguished from random behaviour not only by its direction, i.e., to the future, but also by its strength. In this respect, purpose is like those hold-all terms 'motivation' or 'interest'. Indeed, an objective can only be pursued when one has enough interest in it. Now, it has been said that to sleep is to have lost interest. Wakefulness is then not only a state of consciousness, but also one of more than minimal purposivity, of being interested in whatever one is doing. Whereas boredom is a state in which purposive intensity is low. Such a state, on the border between wakefulness and sleep, must naturally pervade all psycho-physiological parameters and so may make purposivity amenable to research.

Ultradian rhythms

I said above that behaviour may reflect intentional representations. In the endeavour to find some rulefulness in observed behaviour during repetitive and other vigilance-type work, I had often noticed the periodic fluctuations in performance reported by Murrell (1962). I then saw on our filmed records occasions of posture changes and meaningless (non-purposive) behaviour even when no specific work was performed. Apart from the expected instances of 'micro-sleep' during long spells of relative inactivity, sudden episodes of hyperactivity occurred at intervals of approx 90–100 min. The concurrent protocols of subjective reports showed that operators had suddenly become aware of having 'drifted off' and were quickly checking their equipment to reassure themselves. In some cases, we called these episodes 'mini-panics'.

The timing of these incidents led me to consider a possible influence of ultradian, as distinct from circadian, rhythms not only on work performance but also on the general 'stream of consciousness'. What it is exactly that fluctuates rhythmically is not at all clear but, once alerted to their existence, one can find scattered through the literature a great deal of evidence for these variations in physiological functions. Oddly enough, most of the studies are in sleep research. Over 20 years ago, Kleitman (1963) proposed the existence of a Basic Rest Activity Cycle (BRAC), responsible for ultradian rhythms, of about 90 min. These rhythms have since been detected, not only in sleep (i.e., REM-NREM cycles), but also during wakefulness in variables as diverse as oral activity, psychomotor performance, fantasy, waking gastric activity, resting heart rate, eye movements and general alertness (for a review see Hume, 1983). Especially the studies of general alertness (Okawa *et al.*, 1984) and waking fantasy (Kripke and Sonnenschein, 1978) are of interest here in view of our concern for vigilance while the operator is relatively inactive. Some researchers into these rhythms, especially in the laboratory, expect them to recur punctually in exact periodicity. They seem to forget common experience in real life when any vigorous self-activity, like washing one's face with cold water in the morning, breaks the rhythm and postpones the onset of drowsiness, only to re-emerge shortly again. Anyone who has driven a slow car on a motorway for some hours will know that frequent reports of periodic daydreaming, mind-wandering and nodding-off are true, however difficult to record or measure they may be. There can be little doubt that the experience is stressful, especially when the operator is aware of being responsible for other persons. Thus, while responsibility adds another dimension to stress through mental overload, in cases of boredom that same responsibility may mask the effects of a natural body rhythm, if only for a time.

I am convinced that a source of much stress is the inability of even the most responsible operator to anticipate all the errors he knows will probably occur sometime. In what ways can we help him? At least two ways occur to me. One is to extend the study of the common factors and synchronies underlying the variables mentioned above. I do not wish to suggest that Ergonomists should specialize in chronobiology; they might be tempted to learn more and more about less and less. Rather, we should not let go of the puzzling but hidden connections between apparently diverse phenomena like finger tremor and untoward events (Nicholson *et al.*, 1970) or involuntary eye movements, postures and consciousness (Branton, 1985).

The second way in which we may be able to help is an educational-cum-public relations task. Take for example the results of studies of shiftwork and of the effects of circadian and ultradian rhythms on the body and on performance. The task is to disseminate knowledge about human functions and limits to all those people who might possibly be concerned. The critical times and conditions in which human errors and failures are most likely to occur should no longer be esoteric knowledge but should be spread as widely and forcibly as we know how. In such cases, safety considerations must override cost considerations. Sometimes this may do no more than underpin general common sense and experience. If prospective passengers and managers knew that between midnight and 3 a.m. is not a good time to travel by long distance coach there would very likely be fewer coach accidents. Or operators at nuclear power plants might not choose to run experiments on Friday night at 1 a.m. Also, office workers, their managements and everybody else would then know that about an hour and a half without a break is the natural limit of boring work for which a person can reasonably sit without too many costly errors.

Stress in the mental space behind the screen

Unlike any previous work situation, human interaction with electronic workplaces tends to sharpen whatever conflict or incompatibility may exist between the human controller and the artefacts. The processes to be controlled are often many steps removed from his/her direct experience and the very remoteness from the end product generates strong feelings of helplessness, a condition often reported in the literature. No doubt then, the stress of the future will be that connected with VDT work of many kinds. It is upon us already in the form of eye strain, skin rashes, back pain and gynaecological complaints, presumably due to radiation. The epidemiological studies to determine how substantive these are will take at least another five years. But my concern, as might by now be expected, is again with the ergonomics of the hidden factors, rather than the more overt, relatively tangible ones just listed. In part these factors are about:

what is shown on the screen,
what the systems designer intended to represent,
what the operator thinks is the meaning of the display, and
what actually happens 'out there' in the real world which the screen
supposedly represents.

Findings about VDT stress have often been unjustly condemnatory. I think
they make for confusion because of the basic distinctions in mental operations
between dialogue and data entry modes. Each has different implications for
health, wellbeing, safety and performance. The conversational mode deploys the
VDT as a creative aid, where the user is innovator, decision-maker and
controller with some discretion. In these functions, VDTs, keyboards and
accessories appear to exert a strangely compulsive influence on users, comparable to the fascination noted by Postman (1986). The strangeness is in the
conflict with the normally assumed desire for autonomy inherent in people. Not
only does the victim of an obsession with VDT work surrender to unreason, like
gamblers to one-armed bandits, it may also drive programmers and 'hackers'
dangerously close to a kind of unconcern which could end in anarchy. Anyone
involved in the formulation and introduction of international standards can
testify to the state of chaos reigning in this field.

As recent history, e.g., the *Challenger* Space Shuttle, makes abundantly clear,
large groups of computer users can quite unwittingly become prey to addictions
and illusions, subjects to a profound process of divorce from reality. *Challenger*
blew apart before the eyes of millions of viewers, who saw rows of incredulous
and helpless 'controllers' at Houston Mission Control Centre. This side of VDT
work presents to the Ergonomist an entirely unexplored social-psychological
environment (Branton and Shipley, 1986). If new environments create new
forms of life, will future interaction between man and these machines create a
new species – Houston Man? Is not his VDT screen presenting an illusory
picture of the true condition of any plant or process deemed to be under control?
Glued to his screen, how far can he actually control the reality out there?

In vast contrast to the dialogue mode, the application of VDTs to data entry,
increasingly widespread in industry, commerce and finance, currently raises the
majority of complaints referred to earlier. We easily forget that, every time we
write a cheque or use a credit card, some data entry clerk has to input numbers
into the system. This is the Tayloristic conveyor-belt in a new guise and it
presents workload and monotony problems of its own in the new technology
sweatshops of the latest wave of industrialization. In most cases, speed and
variety of electronic information displays create a sharp divide into those who
control and those who are controlled, those who have exciting jobs and those
whose tasks are utterly boring. Controllers and decision-makers are supposedly
aware of the complex consequences of their every action, whilst their 'slaves in
the network' have strictly limited discretion. The former are said to suffer from
mental overload, the latter from 'underload'. Yet both can experience sufficient
stress to incapacitate their performance and perhaps their wellbeing. A common

strategy is to limit their exposure by manipulation of rest periods, but that does not get to the root source of the hazard and perpetuates badly-designed jobs. We are at present investigating this type of work in the laboratory.

There is, however, no need to be gloomy about the future of VDT work. In the long run, 'Workstations shall speak unto workstations across any distance' as has been said recently. Whether whole city populations will be dispersed or not, one thing is certain: things will be quite different, and unexpectedly so. What will be greatly enlarged is the 'mental space behind the screen'. It is for Ergonomists to see to it that the hidden representations correspond as closely as possible to publicly accessible reality.

Postscript

I have just found out that the term Ergonomics is older than English-speaking human factors persons think. Apparently it was first coined as a new word in 1857 by Jastrzebowski in Poland (Laurig, 1982). For its 130th anniversary we therefore need not apologize for re-introducing a word, the meaning of which is not easily definable. To try, nevertheless, I offer the following: If the research and theoretical side can be defined as

the systematic study of the reality of limitations to interaction between humans and their work environment and with each other at work,

the other – practical – side is

the application of that reality to design and use of all kinds of artefact.

It is hoped that the foregoing will have shown the necessity of such a discipline without presuming to formulate ready-made principles. If and when the ideas expressed here are seen as growing points of new theories my enthusiasm may be justified.

References

Akerblom, B., 1948, *Standing and Sitting Posture*, Stockholm: AB Nordiska Bokhandeln.

Bainbridge, L., 1983, Ironies of automation. *Automatica*, **19**, 775–779.

Bainbridge, L., 1985, Inferring from verbal reports to cognitive processes, in *The Research Interview*, pp. 201–215. London: Academic Press.

Blakemore, C., 1976, Feature detection: the concept of perceptual features, in *International Encyclopedia of Psychiatry, Psychology, Psychoanalysis and Neurology*, Wollmann B., (ed), New York: Macmillan.

Branton, P., 1969a, An Ergonomic Study of Factors Affecting the Comfort of Easy-Chairs. Unpublished MPhil thesis, University of London.

Branton, P., 1969b, Behaviour, body mechanics and discomfort, *Ergonomics*, **12**, 316–327.

Branton, P., 1978, The train driver, in *The Study of Real Skills, Vol.1: The Analysis of Practical Skills*, Ch.8 Singleton W. T. (ed) Lancaster: MTP Press.

Branton, P., 1979, Investigations into the skills of train-driving, *Ergonomics*, **22**, 155–164.

Branton, P., 1982. The use of critique in meta-psychology, *Ratio*, **24**, 1–11.

Branton, P., 1984, Transport operators as responsible persons in stressful situations, in *Breakdown in Human Adaptation to 'Stress'*, Vol. 1, Part 2, pp. 494–508, Wegmann H.M. (Ed.) Dordrecht, Netherlands: Nijhoff, for Commission of the European Communities.

Branton, P., 1985, Some epistemological consequences of recent psycho-physiological discoveries. Paper presented to XI Inter-American Philosophy Congress, Guadalajara, Mexico, November, 1985.

Branton, P. and Grayson, G., 1967, An evaluation of train seats by observation of sitting behaviour, *Ergonomics*, **10**, 35–51.

Branton, P. and Oborne, D., 1979, A behavioural study of anaesthetists at work. In *Psychology and Medicine*, pp. 434–441, edited by Oborne, D. Gruneberg M. and Eiser J. R., (Eds), London: Academic Press.

Branton, P. and Shipley, P., 1986, VDU stress: is 'Houston Man' addicted, bored or a mystic?, in *Proceedings of International Scientific Conference: Work with Display Units*, Stockholm, 1986, Knave B. and Widebaeck. P. (Eds).

Granit, R., 1977, *The Purposive Brain*, London: MIT Press.

Hubel, D. H. and Wiesel, T. N., 1962, Receptive fields, binocular interaction and functional architecture in the cat's visual cortex, *Journal of Physiology*, **160**, 106–154.

Hume, K. I., 1983, The rhythmical nature of sleep, in *Sleep Mechanism and Functions in Humans and Animals: An Evolutionary Perspective*, Mayes A. (Ed), Wokingham: Van Nostrand Reinhold.

Kleitman, N., 1963, *Sleep and Wakefulness*, Chicago: University Press.

Kripke, D. F. and Sonnenschein, D., 1978, A biologic rhythm in waking fantasy, in *The Stream of Consciousness*, pp. 321–332, Pope P. and Singer J. L. (Eds), New York: Plenum Press.

Kuhn, T. S., 1970, *The Structure of Scientific Revolutions*, Chicago: University Press.

Laurig, W., 1982, *Prospektive Ergonomie – Utopie oder Wirklichkeit?*, Cologne: AGV Metall.

Murrell, K. F. H., 1962, Operator variability and its industrial consequence, *International Journal of Production Research*, **1**, 39.

Nelson, L., 1949, The critical method and relation of psychology to philosophy, in *Socratic Method and Critical Philosophy*, pp. 105–157, New Haven: Yale University Press.

Nicholson, A. N., Hill, L. E. Borland, R. G. and Ferres, H., 1970, Activity of the nervous system during the let-down, approach and landing: a study of short duration high workload, *Aerospace Medicine*, **14**, 436–446.

Okawa, M., Matousek, M. and Petersen, I., 1984, Spontaneous vigilance fluctuations in the daytime, *Psychophysiology*, **21**, 207–211.

Polanyi, M., 1958, *Personal Knowledge*, Chicago: University Press.

Popper, K. R., 1983, *Realism and the Aim of Science*, p. 132, London: Hutchinson.

Popper, K. R. and Eccles, J. C., 1977, *The Self and Its Brain: An Argument for Interactionism*, London: Springer.

Postman, N., 1986, *Amusing Ourselves to Death*, London: Heinmann.

Pribram, K. H., 1971, *Languages of the Brain: Experimental Paradoxes and Principles in Neuropsychology*, Englewood, NJ: Prentice-Hall.

Selye, H., 1976, *The Stress of Life*, New York: McGraw Hill.

Shackel, B., Chidsey, K. D. and Shipley, P., 1969, The assessment of chair comfort. *Ergonomics*, **12**, 269–306.

Singleton, W. T., 1972, Techniques for determining the causes of error, *Applied Ergonomics*, **3**, 126–131.
Young, L. R. 1969, On adaptive manual control, *Ergonomics*, **12**, 635–674.

Part IV
An Annotated Bibliography of
Brantonian Publications

Bibliography

The following is a chronological list of Paul Branton's publications. They have been lodged with the Ergonomics Information Analysis Centre, Department of Production Engineering, University of Birmingham (Tel: 021-414 4239) from where copies may be obtained. Publications marked with an asterisk (*) are reproduced in full in Part III of this book.

1. *A model of the central functions of human social conduct.* Unpublished manuscript, 1960.

> This paper was produced by Branton while still a student at the University of Reading. It represents the early stages of the development of the Brantonian View and considers the purposes of the individual when engaging in social conduct. Within the paper he develops ideas about the nature of social interactions and the knowledge which they are intended to gather. He discusses the importance of uncertainty reduction, for example, and the stimulus seeking behaviour which can help towards it. Even though Branton asserts that: 'the model will concern itself with what goes on in one human when he interacts with another human; what happens when the human interacts with anything non-human will be considered as a fundamentally different matter' (p2), many of the principles he expounds, including drawing on physiological evidence, may be seen as the early stages of the Brantonian View discussed in this book.

2. Seating in industry, *Ergonomics for Industry*, **10**, London: Ministry of Technology, 1966.

> The theme of this small pamphlet develops strongly the need to design for function and performance, rather than merely to fit the operator's physiological dimensions. It considers the important measurements in relation to the design of seats for the sitter. The purpose of seating is discussed, as is the relationship between posture and (both sitter and seat) performance.

3. The comfort of easy chairs: an interim report on the present state of knowledge, 1966, *FIRA Technical Report* **22**, Stevenage: The Furniture Industry Research Association.

> This report adopts a decidedly person-centred approach to chair design and to the nature of comfort. It takes designers to task, for example, for advocating what Branton calls the 'moral' approach to design: 'Most researchers first decide on an idealised, "desirable" sitting posture, usually on the grounds of health, and then

223

proceed to design a seat to achieve that posture. In effect they say, "This is how you *should* sit".' Branton, on the other hand, argues for a more naturalistic, behavioural, approach to sitting and seat design: '. . . seats are not sat on in only one posture, but in many ways. The sitter is a living being, is usually doing something whilst sitting and it is he, not the chair, who will be comfortable or not.'

4. School furniture dimensions: standing and reaching, 1967, *Building Bulletin*, **38**, London: Department of Education and Science, HMSO.

This is a report of a study carried out in October 1965 by Branton while an experimental officer at FIRA, and commissioned by the Department of Education and Science. Although the report provides a considerable number of dimensions appropriate for the design of school office furniture, it also emphasises that designers should consider the design in terms of the children's *performance* rather than their body part measurements. The instructions to children given during the trials illustrate the importance placed on this emphasis, as well as provide interesting illustrations as to how a person-centred view can be accomplished easily.

5. An evaluation of train seats by observation of sitting behaviour. (With Grayson, G.), 1967, *Ergonomics*, **10**, 35–51.

Before he joined British Rail as Chief Ergonomist, Branton was asked to evaluate two different prototype seats for future railway carriages. Rather than simply asking sitters to judge the 'comfort' or other attributes of the seats, he introduced a more naturalistic study based on his assertion that the postures which people adopt provide an ideal 'window' on the person's inner feelings. By analysing sitter's postures over a five hour train journey, he was able to demonstrate that the two seats could be differentiated on the basis of frequency of occurrence, duration and sequences of postures. Furthermore, by using speeeded-up filming techniques which made otherwise imperceptible postural changes perceptible and thus analysable, he was able to demonstrate that one seat produced less stability than the other when in a dynamic environment like a moving train.

6. A note on the relation between self-rating scales and other measures. (With de la Mare, G. and Walker, J.), 1969, *Occupational Psychology*, **43**, 316–327.

This note reports a small study which demonstrated that self ratings do not generally correlate well with objective measures of the factors underlying the ratings.

7. *Behaviour, body mechanics and discomfort, 1969, *Ergonomics*, **12**, 316–327.

In this review Branton develops his argument that the definition of comfort is an elusive aim: 'observation of verbal and postural behaviour are then to be interpreted in terms, not of the experience of comfort, but of motivation to avoid interference with primary activities, or of avoiding *dis*comfort . . . a seat may be measurably inefficient to the degree to which it interferes with the primary activity'. Thus the argument is developed that ergonomists should consider the individual's motivations (in this case for sitting) as a means of measuring such 'internal' feelings. Branton also uses the review to develop his theory of postural homoeostasis in which the seat needs to accommodate the competing demands for stability while, at the same time, allowing the sitter the freedom of movement necessary to fidget and relieve pressure.

8. The user in transport, 1969, *RIDE Conference Paper*, Institution of Mechanical Engineers.

The importance of the user's viewpoint is centrally placed within this paper for railway engineers. As an ergonomist, Branton argues that the railway environment is so poor because 'we still take the user too much for granted. There is an *imbalance* between the technical achievements of railways and *the use of them.*' The person-centred views of purposivity, control and anticipation become important. Discussing the design of railway timetables, for example, Branton asks: 'What does the user look for? He only knows his ultimate destination. Why then should timetables for the public be linear? or chronological?' More importantly than specific facets of railway design, Branton felt, design questions should be asked from a user-centred view: 'How will *naive persons* find their way? How and when do they get to the *right station*? What do they do with their luggage? *How much* luggage have they?'

9. Review of Furniture and equipment dimensions: further and higher education, 18–25 age group, *DES Building Bulletin* **44**, London, HMSO, in, *The Architect's Journal*, 8 July 1970.

Although it discusses a very specific aspect of design, this short book review concludes with a person-centred plea for ergonomics: 'Let teachers and children work in an environment designed to give and encourage freedom of movement and expression'.

10. A field study of repetitive manual work in relation to accidents at the work place, 1970, *The International Journal of Production Research*, **8**, 93–107.

Four hundred and twenty-seven reported accidents in the course of repetitive, self-paced work in the machine shop of a light engineering factory were analysed for time of occurrence. Four critical peak periods were found. During these periods observational studies of variability of speed of operation were conducted, particularly on lathes, which revealed that machine loading times varied more than cutting times. This was followed by a case study of variability of accuracy of hand movements. Unsuccessful Hand Movements (UHMs) were found to occur more often during the critical periods than at other times of the day. Their data are interpreted in terms of rate of gain of information, fatigue and boredom. (Journal Abstract)

11. **Train drivers' attentional states and the design of driving cabins.* Paper presented to 13th Congress, Union Internationale des Services Medicaux des Chemins de Fer, Ergonomics Section, Brussels, October 1970. Published in the Proceedings.

This paper develops many of the person-centred principles discussed within this book. The nature of the operator as an active doer, as opposed to a passive responder, is posited, as well as the nature of the activities in which the individual engages. The paper also introduces the importance and the nature of many of the physiological rhythms which underlies these activities. The importance of 'Design from the man out' is a central strand of the argument produced.

12. Seating in Industry, 1970, *Applied Ergonomics*, **1**, 159–165.

This article is a copy of the pamphlet with the same title published by HMSO (1966).

13. *Ergonomic research contributions to design of the passenger environment, 1972, in *Passenger Environment*, 64–69, Proceedings of a conference organised by the Institution of Mechanical Engineers.

This paper discusses the role of methodology in ergonomics and concentrates on three main areas for consideration. The first concerns the difference between answers to questions about measurable facts and actual measurement. The second deals with subjective comfort and the difficulties of describing it. The third describes attempts at demonstrating the existence of consistency in the subjective feelings evoked by environmental design. Throughout the paper Branton enlarges on the complexities involved in asking people questions and the reliance to be placed on the answers – in other words, when one tries to turn subjective statements and judgements into objective evidence.

14. Ergonomics, 1973, in *Occupational Health Practice*, Schilling, R.S.E., (Ed), London: Butterworth.

This chapter provides a general overview of the basic tenets of traditional ergonomics, though without the mechanistic view which tradition presents. Topics include the nature of the person at work, in terms of anthropometric considerations, information transmission, and energy expenditure. Also covered are some of the motivations which people bring to work and postural aspects of comfort-seeking behaviour. Methods of data collection are discussed, including measures of speed and accuracy, 'error' measurement, psychophysical and subjective scaling, and physiological measures. The chapter concludes with a consideration of the person being part of a cybernetic system at work, in which the concept of 'error' as departure from a desired path is considered. Throughout, Branton argues strongly for a philosophy of ergonomics which places people at the centre of the system and designs 'from the man out'.

15. *Anthropometry and the design of internationally usable technology*, unpublished paper, 1973.

This manuscript was produced originally for publication in *Human Factors*. It argues that it is extremely important for ergonomists to recognise the existence of anthropometric differences between people of different ethnic origins when designing equipment to fit the individual. Otherwise the 'natural adaptability' of people to less than optimum working conditions may not be recognised, with resultant reductions in well-being and efficiency. The importance of anthropometric considerations does not just relate to the design for static and even dynamic behaviour, however. Branton also includes the importance of designing for function.

16. Review of *Allgemeine Arbeits- und Ingenieurpsychologie: psychische Struktur und Regulation von Arbeitstätigkeiten*, Hacker, W. Berlin: VEB, 1974, in *Occupational Psychology*, 1974.

This review of a book by a German author clearly illustrates Branton's disdain for work-based philosophies which do not recognise the pre-eminent role of the person within the system. The book being reviewed took a decidedly Marxist-Leninist approach to using work as a way of maintaining social order. Despite his concern at the somewhat overbearing approach within the book, Branton concludes that the development of individual personality at work may well help to produce a new social order if it were to be 'given true scope and freedom'. Within this review, therefore, the Brantonian View begins to emerge clearly.

17. Review of *School furniture: standing and sitting postures*, DES Building Bulletin 52, London, HMSO, 1974, in *The Architects' Journal*, 1974.

The encouragement for the development of a person-centred ergonomics philosophy continues with this book review. While realising that it would be unrealistic to expect to design a different seat to fit each individual school child, Branton praises the book's authors for emphasising a scientific approach to understanding the range of body structures and task functions for which the chairs will be used. The review concludes with a plea to 'let teachers and children work in an environment designed to give and encourage freedom of movement and expression'. Clearly, the person-centred principles expounded within this book are encapsulated in this plea.

18. An experimental (ergonomic) evaluation of prototype velocity displays for the BR Advanced Passenger Train. (With P. Shipley). *IEE Conference Publication No 150*, 1977 *Displays for man-machine systems*, London: Institute of Electrical Engineers.

This paper reports a study to evaluate different kinds of speed displays (analogue and digital) for use in the train driver's cab. A variety of ergonomics recommendations resulted, particularly in relation to the design and positioning of digital displays among others on the console. As far as the development of a person-centred approach is concerned, the authors point out that the train driver continually has to predict the likely course of events and that the displays provided should enhance these tasks: 'the driver will be posing to himself questions about his actual speed, its present trend, and his required speed, constantly estimating the direction and magnitude of any discrepancies . . . The driver spends most of his time monitoring and processing trackside information through his cab windows, and can afford little time switching attention to displays inside the cab.'

19. Choosing chairs, 1977, *Talk Back*, 3, 4–6.

In this newsletter article Branton expounds the importance of realising the unselfconscious aspects of behaviour – in this case sitting behaviour. Sitting, he argues, is a naturally spontaneous activity; the postures adopted by the sitter will be such that the 'ideal' sitting posture will be taken, almost 'unconsciously'. He develops five principles: sitting is un*self*conscious, it is an activity, it is a means to another end (i.e. it has purpose), the postures adopted are defined by the body structure, and sitting behaviour is developed so that the body is at the same time stabilized and allowed to be mobile.

20. **On the process of abstraction*, 1977 unpublished paper.

In this paper I shall first describe what is often regarded as the course of abstraction. I shall use an example from Karl Pearson's *Grammar of Science* and then discuss the representation of 'concepts' in mind/brain. Concepts, I shall argue, are percepts idealised in the process of internal representation. This will lead me to try and specify what I mean by the internal representation of an 'ideal' and an elaboration of the mutual relation between ideals and success in skilled acts. (Author's abstract)

21. The train driver, in *The Study of Real Skills. Vol I: The analysis of practical skills*, 1978, Singleton, W. T., (Ed), Lancaster: MTP Press.

In this chapter Branton expands the person-centred theme to discuss the nature of train driving skills and the extent to which they are the product of the autonomous, responsible operator working in a quasi-mathematical way to predict the future course of events. Driving a train is not an 'automatic process', nor one which can be mechanised in any robotic way. Rather, the skill involves integration of current knowledge with past experience and future predictions. The train driver 'acts in an isolated, mobile environment, dependent on long-delayed feedback with a unidimensional output, unlike that of air, sea and road vehicle controllers'. Branton concludes in a person-centred vein by arguing that 'if his [the train driver's] skill is deliberately built into the system, given equipment well fitted to his own limitations, and if he is allowed to operate in an environment moderately and appropriately enriched so that he can again 'feel at one with the train', he will do the rest'.

22. *Testing and evaluation of transport drivers seats, a state-of-the-art report*. Commissioned by the National Swedish Road and Traffic Research Institute, 1978.

This review deliberately centres attention on seats as integral parts of truck drivers' work stations. It is assumed that better understanding of the driver's functions and needs for suitable suspension and support leads directly to prevention of discomfort and ill health, and contributes to road safety by enhancing skilled performance. There is, however, no generally accepted standard procedure for evaluating seats because human functions and driving skills are not well enough understood. The history of attempts at applying the ISO vibration standard underlies the need to clear the purpose and meaning of the test before writing their mechanical specifications. (Review abstract)

23. *Investigations into the skills of train driving, 1979, *Ergonomics*, **22**, 155–164.

Train driving as a control task is one-dimensional, yet very complex. This combination highlights certain problems in our understanding of skilled action. Investigations involving behavioural observations, plus interviews with over 200 drivers and inspectors, showed that the drivers utilise more information from outside the cab than is usually thought. The relevant variables were identified. The limitations to the driver's possible knowledge of the changing state of the system ahead of him lead the study to the goal-directed, purposive nature of his skill. What exactly does he have to carry in his head to achieve the observed successes in time-keeping and safety? Consideration is given to the form of internal representations of his outside world. Quasi-mathematical operations to solve time/distance trajectory equations are suggested. Enactive, rather than verbalised, storage of information is discussed. Some practical consequences for training and equipment design are drawn in conclusion. (Journal abstract)

24. A behavioural study of anaesthetists at work, (With D. J. Oborne), 1979, in *Research in Psychology and Medicine* (Eds Oborne, D. J., Gruneberg, M. M., and Eiser, J. R., London: Academic Press.

This paper reports a video-recording study of anaesthetists which incorporated psychophysiological measures including air sampling, ECG, and eye-blink rate recording. The results indicated a fall in anaesthetist heart rate as the operation progressed, though with peaks in psychophysiological response at periods of operational stress. Behavioural measures demonstrated the occurrence of 'mini-panics' during the operation – 'sudden transient states of excitement and possibly adrenal flooding – as the anaesthetist struggles with drowsiness (and the realisation that drowsiness has occurred) due to 'self-imposed inactivity and postural immobility'.

25. The research scientist, in *The Study of Real Skills. Vol II: Compliance and Excellence*, 1979, Singleton, W. T., (Ed) Lancaster: MTP Press.

In this chapter Branton applies a person-centred analysis to another kind of skill: that of the research scientist. The technique which he reports is to interview a number of eminent scientists in an attempt to abstract from the protocols the basic characteristics which led to their success. Scientific research, he argued, represents a skill in which the process involves work which the individual has never done before; the researcher's excellence is a measure of his or her ability to extend the 'self', not simply 'competition' with others. Thus the skill contains characteristics of thought and problem solving, experience, intuition, and abstraction. It also includes communicative abilities insofar as the successful researcher must be able to communicate results and ideas with others. Couched in such terms, therefore, Branton argues for a person-centred analysis of the scientific researcher, rather than an analysis which is based simply on the outcomes of the research.

26. The concept of comfort and its measurement, 1980, *Nursing*, No. 20, 856–857.

'The "grammar" of comfort is not yet well established and is as much an art as a science.' The statement sets the scene for this paper, which considers all aspects of the nature of comfort: its measurement, planning and dimensions. The paper begins to argue that 'comfort' itself is a misnomer, that people can really only distinguish between levels of '*dis*comfort', and that the subjective response to external factors which may lead to such effects can be ameliorated by person-centred attributes such as motivation and attention.

27. *A psychology of reasonable autonomy*, 1981, unpublished manuscript.

This is Branton's most extended attempt at a synthesis of his ideas. After a summary (chapter 1), he begins with two questions of methodology (chapter 2): the rejection of all question about origins as speculative and scientifically unknowable, and the acceptance as inevitable of some concept of 'faculty' or human capacity, whose excesses we may face by means of a sophisticated new brand of operational description. (He shows that all psychologists use a version of 'faculties', but try to hide the fact beneath a new terminology.) Branton also introduces here the concept of a 'psychological quantum jump' – the most basic inner change taking place in the head – insisting that all psychological processes (all becoming aware of something, all getting-to-know, all deciding-to-act, and lastly all acting) have in fact this mysterious jump-like nature. After applying the 'psychological quantum' concept to visual perception, Branton moves on to reasoning as an operation basic to all action (chapter 3). He analyses reasoning into four human capacities corresponding to four well-established facts of human cognition: that attention is selective, that a process of abstraction occurs, that features of objects are preserved in the head, and that representations allow for vicarious manipulation of objects, particularly with regard to future possibilities. In order for action to actually occur, however, a fifth factor is required – conferring of values to things and events. This is the subject of a separate chapter (chapter 4), in which Branton shows how values arise out of human needs, how all human life is given continuity by moods, how all feeling contains a hidden judgement of value, and how social skills are made possible by such needs, moods and feelings. Thus the way is paved to the centerpiece of the whole book: Branton's analysis of moral (or ethical) action, into five steps, each one of which has a corresponding basic human capacity: sympathy, empathy, abstraction, weighing-up and wilful decision (chapter 5). There is a 'psychological quantum threshold' dividing socially skilled acts where no conflict of interests occurs and ethical actions proper where one has to choose a course of action, although the eventual decision may be painful for oneself or others. In Chapter 6 Branton concentrates on the last step – the much vexed problem of the will – trying to show how volition can be amenable to scientific treatment. Chapter 7 is mainly methodological, regarding the need to replace causal by purposive explanations in psychology. Given the importance and difficulty of this topic, it is no wonder that Branton made several attempts at rephrasing his basic insights on purposivity later in his life (1981, 1983, 1985, 1989). In Chapter 8, which is also methodological, he tries to sketch a new psychology which would follow the lines indicated in the book. It is his most systematic and interdisciplinary chapter, and it contains brilliant insights. The last chapter (chapter 9) is an extension of the argument of Chapter 8 to larger philosophical questions. (Abstract by F. Leal)

28. *The use of critique in meta-psychology, 1981, *Ratio*, **24**, 1–11.

In this paper, written in the year of Leonard Nelson's 100th birthday, Branton expands his views concerning the important links between philosophy, particularly The Critical Philosophy, and psychology. He discusses the nature of basic mental facilities and the extent to which they enable pre-conceptual thought to occur. It is in this paper that Branton considers the relationship between thinking and doing, and the nature of abstractional thought. The concept of the individual as a reasonable being, and the influence which such thought has on the nature of active behaviour, is also discussed.

29. *Operators as reasonable persons – how to optimise their reliability*, paper presented to Pilkington seminar, 29 June, 1982.

The importance of realising the responsible nature of people at work is highlighted in this short presentation. In particular, the problem of underload and boredom in less than optimally stimulating environments is discussed. Emphasis is placed on the operator's own realisation that the lack of stimulation is likely to lead to reduced attention, and the 'mini-panics' which may result: 'Reliability is as much a matter of the operators' intelligent self control as of any attempts by others at controlling and managing them. They must be given a chance to exert themselves and anticipate responsibly their own mental work situations. If they know about their own nature, how liable they naturally are to making mistakes, to mind-wandering, day-dreaming, to introspection and to twilight states or to dozing off, and roughly when this might happen, they may be able to take preventative action themselves.

30. *On being reasonable, (Vernunftig sein), 1983, in *Vernunft, Ethik, Politik: Gustav Heckmann zum 85. Geburtstag.* Horster, D. and Krohn, D., (Eds) Hannover: SOAK Verlag.

Branton's person-centred view of behaviour is strongly apparent in this paper. The concept of the reasonable, thoughtful individual who behaves in a responsible manner is emphasised; the implication for ergonomics being that environmental design needs to facilitate such behaviour and not expect people at work to operate in almost automaton ways. The paper begins with the view: 'At a time when the world seems to have gone crazy, the cry from some quarters is "Let us be reasonable – not coldly, calculatingly rational – just reasonable". Being reasonable is, above all, being thoughtful, in contrast to being impulsive. The thoughts are, however, not just idle ones but thoughts with a view to action, practical rather than merely contemplative.' The remainder of the paper then discusses in detail the nature of this reasonable behaviour.

31. *Process control operators as responsible persons. Invited paper to symposium on *Human Reliability in the Process Control Centre*, Institution of Chemical Engineers, Manchester, April, 1983.

From reasonableness to responsibility. This presentation discusses the nature of the responsible person at work – in this case process control operators who normally have to perform continuous – often monotonous – tasks. In such environments mind wanderings, reduced interest, and mini-panics can occur – all of which can be stressful for the operator. In this paper Branton also includes consideration of the body rhythms which may accompany such behaviours.

32. *Critique and explanation in psychology – or how to overcome philosophobia*, 1983, unpublished paper.

This paper provides a valuable insight into Branton's view of the value of Critique in the development of psychological investigations. 'The first thing to be said for critique is that it encourages the habitual focusing on what questions are to be asked before rushing out of the lab again for further evidence. . . . An even more fruitful aspect . . . is that . . . it asks about the *necessary and sufficient knowledge* without which the persons observed could not have behaved as they actually did.'

33. Backshapes of seated persons – how close can the interface be designed?, 1984, *Applied Ergonomics*, **15**, 105–107.

The backshapes of 114 seated persons were measured. The horizontal and vertical positions of various spinal landmarks are given for male and female subjects and for those with lordotic, straight and kyphotic backs. (Journal abstract)

34. *A cognitive function of rapid eye movements*, paper presented to the Symposium of the European Society for the Study of Cognitive Systems. Cambridge, June, 1984.

This paper brings together three separate theoretical questions of perception and action. One is pre-perceptual: How can one possibly see a stable world when one's eyeballs are never still? The second is: What can be represented internally? The third asks: What could be the purpose of eye movements in dreams, when conscious sensory input is almost wholly suppressed? Branton argues that Rapid Eye Movements (REMs) have a function beyond retinal reception and that the processes involved are so complex and inaccessible that more than one kind of explanation may be necessary to over-arch phenomena ranging from simply seeing objects to dreaming and fantasising, remembering and reporting about them and using them in reality. (Paper abstract).

35. *Transport operators as responsible persons in stressful situations, 1984, in *Breakdown in human adaptation to stress: Towards a disciplinary approach. Vol I.* Wegmann, H. M. (Ed), Boston: Martinus Nijoff.

This paper returns to the person-centred argument which views people in controlling operations as being susceptible to the competing demands of boredom or fatigue and the necessary expression of responsibility at work. The nature of the competition is such as to increase mental load and stress. Branton also considers different ways of conceptualising such tensions and their implications. These include the 'engine model', a mechanistic view of behaviour which considers just the input/output relationships of the interaction; the 'organic model', which looks more to the activities of the operator's internal body mechanisms to explain events; the 'personality differences' model, which explains events in terms of attitudinal or motivational factors; and the 'information processing' model, which views the interaction in terms of consciously cognitive decisions. Branton rejects each of these models in favour of a model 'of the operator as a goal-directed controller, autonomous, yet liable to natural fluctuations in consciousness'.

36. *Some epistemic consequences of recent psycho-physiological discoveries*, paper presented to the 11th Philosophical Inter-American Congress. Guadalajara, Mexico, November, 1985.

In this paper, Branton draws together psychophysiological findings with philosophical observations to begin to answer questions relating to the knowledge representations which people develop when interacting with the world. His psycho-physiological references deal particularly with rapid eye movements and the progress of perceptual events from the retina to the visual cortex, and his argument relates such information to understanding of the Self.

37. The social importance of ergonomics, 1985, *The Ergonomist*, **183**, 2–3.

This paper makes explicit the central tenet of person-centred ergonomics: 'Persons are fallible, error prone and easily bored; but – *most importantly* – only they are capable of making valid choices and decisions'. The 'social importance' to which Branton alludes in his title concerns the need for more aspects of our lives, beyond ergonomics, to adopt a person-centred view of interactions – social as well as physical. However, it is up to ergonomics to show the way: 'Application of ergonomic principles would save much trouble and release people to contribute greatly to national wealth and social life'.

38. Visual presentation of EEG frequency patterns as a means of gaining a measure of personal control over electrocortical activity, paper presented to CEC workshop *Electroencephalography in Transport Operations*, Cologne, September, 1985. Published in proceedings Gundel A., (Ed).

This is a preliminary report on a portable instrument which displays EEG frequency patterns visually while simultaneously allowing a tape recording for later replay and quantitative analysis. Originally designed to monitor patients undergoing stress therapy, the instrument is used for relaxation biofeedback; its use as a heuristic tool is demonstrated here. One of the pressing problems of transport operators and others working in monotonous environments is to overcome stress arising from awareness of being under-aroused. In the first part of the paper, a pilot study is described briefly as background to the second part, a discussion of methods and purposes of further research into the control over drowsiness and sleep. (Paper abstract)

39. *VDU Stress: Is Houston Man addicted, bored or mystic? (With Shipley, P. 1986, in *Proceedings of International Scientific Conference on Work with Display Units*, Knave, B. and Widebaeck, P. (Eds), Stockholm, May, 1986.

This paper discusses the 'social-psychological' environment which continuous working with VDUs may engender. It posits that working through VDUs creates a situation in which the processes being controlled are 'at least one step removed from the controller's direct experience'. To what extent can this step be trusted? And how may this almost divorce from immediate reality lead to operator stress? The implications of such questions, of course, lead us to consider ways in which the hardware and software inherent in VDU technology may be designed to reduce any feelings of divorce from reality.

40. *In praise of ergonomics – a personal perspective, 1987, *International Reviews of Ergonomics*, **1**, 1–20.

This paper describes, in a personal way, a search for the links which could possibly bind together the many, apparently quite disparate, areas to which Ergonomists devote their efforts. Naturalistic observation of people's behaviour highlights some puzzles that cry out for explanation. There is, for instance, their inability to verbalise even their own most simple skilled actions. To perform as they do, their world must have been represented internally in certain forms. Yet to be explained by neuropsychology, that world must be replicated in all its complexity by continuous processes in neural (brain) tissue, and the mental operations upon these representations must be both extremely precise and yet flexible enough to foresee future states of the man-machine interactive system. It is shown that valid explanations must go beyond current preoccupations with computer analogies. The explanatory power of the concepts of autonomy and purposivity are explored. A broad concern with principles is necessary if responsible Ergonomists are to provide help in overcoming the hazards of Chernobyl and the Space Shuttle. Ergonomics, with its emphasis on practical application to normal humans, is seen as the life science par excellence. (Journal abstract)

41. *Causes or reasons as determinants of deliberate acts*, 1989, unpublished paper responding to Grete Henry-Hermann 'Conquering Chance', published in *Philosophical Investigations*, 1991, **14**.

This critique begins from Henry-Hermann's question, posed originally in 1953, whether actions may have no cause at all. The distinction is made between causes (mechanistic precursors of events) and reasons (which are more subjective and relate also to purpose). Branton goes on to consider the nature of internal representations of the outside world and the extent to which they allow time for the person to reason and to consider the putative consequences of actions *before* they become public. He argues that this suggests the existence of a kind of knowledge which is non-verbal, pre-conceptual and expressed at all levels of skilled acts. This is best called *pre-formative* knowledge.

Additional publications

A bi-polar measure of autonomic nervous activity, BA Thesis, 1962, University of Reading.

An ergonomic study of factors affecting the comfort of easy chairs, MPhil thesis, 1969, University of London.

Macht und Ohnmacht des Bürgers in userer heutigen Gesellschaft, paper presented to Geist und Tat workshop, Cologne, May, 1977.

Causal and purposive explanations of human behaviour – towards a further development of Nelson's practical philosophy, (In German), paper presented to the Philosophical-Political Academy, Frankfurt, November, 1981.

Index

Person-centred ergonomics